To
Arline H. Bunch
who always expected me to
write a book

Mathematics is often erroneously referred to as the science of common sense.

Edward Kasner and James R. Newman

Mathematics may be defined as the subject in which we never know what we are talking about nor whether what we are saying is true.

Bertrand Russell

There is no permanent place in the world for ugly mathematics.

G. H. Hardy

Cantor was aware of this problem but took it in his stride, which I think was very perceptive of him. Russell also found out about it, and he worried about it.

J. N. Crossley

· · · this was sometime a paradox, but now time gives it proof.

William Shakespeare

Mathematical Fallacies and Paradoxes

Bryan Bunch

DOVER PUBLICATIONS, INC.
Mineola, New York

Bibliographical Note

This Dover edition, first published in 1997, is an unabridged and corrected republication of the work first published by The Van Nostrand Reinhold Company, New York, in 1982.

Library of Congress Cataloging-in-Publication Data

Bunch, Bryan H.
    Mathematical fallacies and paradoxes / Bryan Bunch.
        p.   cm.
    Originally published: New York : Van Nostrand Reinhold Co., c1982.
    ISBN 0-486-29664-4 (pbk.)
    1. Logic, Symbolic and mathematical.   I. Title.
QA9.B847   1997   97–15386
511.3—dc21   CIP

Manufactured in the United States of America
Dover Publications, Inc., 31 East 2nd Street, Mineola, N.Y. 11501

Preface

This book is a collection and analysis of the most interesting paradoxes and fallacies from mathematics, logic, physics, and language. It also treats important results in mathematics that are based in paradox, notably Gödel's theorem of 1931 and decision problems in general.

The material is arranged so that the rather tenuous relationship between mathematical reality and physical reality becomes the subject of the book, while the paradoxes and fallacies are tools for exploring this relationship. As a result, although the material contains a number of topics that are often presented in an anthology format, there is a definite progression from the first chapter to the eighth. It is possible, however, to read most of the individual paradoxes or fallacies in whatever order takes one's fancy.

The first three chapters are largely concerned with examples that today are generally classed as fallacies. As such, they have specific defects in the mathematics, defects upon which all mathematicians are in agreement.

It seems appropriate to encourage the reader to try finding those defects. Therefore, I stop at a point where the presentation of the fallacy is complete and ask: Can You Find the Flaw? I also provide a hint. The remaining chapters deal with topics for which there is no single, accepted explanation, so this feature is dropped in Chapters 4 through 8.

I assume throughout that the reader has some experience with the content of first-year high-school algebra. A few of the results also draw upon parts of high-school geometry. Any mathematics that is needed beyond these levels is developed as a topic in the body of the book. This includes a brief introduction to the basic ideas of complex numbers in Chapter 1; mathematical induction, the notion of the limit of a series, and some ideas from probability in Chapter 2; indirect proof in Chapter 3; and elementary set theory in Chapter 5. These are all necessary to a complete understanding of many of the paradoxes. Even so, there are some mathematical complexities that I have omitted deliberately. For example, the discussion of

Gödel's incompleteness theorem is necessarily simplified (even the type of incompleteness to which it applies is omitted), as is the analysis of special relativity. These simplifications in no way affect the results stated. In a few cases, it has seemed better to state results without proof, rather than to become bogged down in lengthy and difficult mathematical development.

I wish to express my gratitude to Dr. Phillip S. Jones of the University of Michigan for his thoughtful comments on much of the material in the manuscript; and to my wife, Mary, for retyping the whole manuscript, as well as for her help and support in so many other ways.

Briarcliff Manor, NY

B.H.B.

Contents

1
Thinking Wrong about Easy Ideas

One should not forget that the functions, like all mathematical constructions, are only our own creations, and that when the definition with which one begins ceases to make sense, one should not ask, What is, but what is convenient to assume in order that it remains significant.

Karl Friedrich Gauss

Everybody makes mistakes. In particular, everybody makes mistakes in mathematics. "Everybody" includes mathematicians, even some of the greatest of all time.

If you add two numbers and get a wrong sum, the mistake is just a mistake. If the wrong answer results from an argument that seems to make it correct, the mistake is a fallacy. Sometimes students give explanations for mistakes that sound very logical. But the sum is still wrong.

Here is a very simple example of such a mistake. This one is made deliberately. You may have seen the following trick played on a child (or even on an unsuspecting adult). You go to the child and say that you can prove that he or she has 11 fingers. "How?" says the child, who knows perfectly well that almost everyone has 10 fingers. "By counting."

You proceed to count the fingers in the ordinary way, laying one of your fingers on each one of the child's in turn.

$$1, 2, 3, 4, 5, 6, 7, 8, 9, 10$$

"Oh," you say, "I must have made a mistake. Let me do it the other way." Starting with the little finger you count backwards:

$$10, 9, 8, 7, 6$$

You are now at the thumb. You stop and say, "And 5 more from the other hand, which we counted before, makes 11."

An incorrect result coupled with an apparently logical explanation of why the result is correct is a *fallacy*. (The word *fallacy* is also used to refer to incorrect beliefs in general, but in mathematics the incorrect chain of reasoning is essential to the situation.) In fact, the word in mathematical usage could also refer to a correct result obtained by incorrect reasoning, as, for example, the student who used cancellation for reducing fractions as follows:

$$\frac{1\cancel{6}}{\cancel{6}4} = \frac{1}{4}$$

and

$$\frac{1\cancel{9}}{\cancel{9}5} = \frac{1}{5}$$

The student got the correct result in these cases, but the method has no logical basis and generally would fail.

Sometimes the mistakes in reasoning come because your experience with one situation causes you to assume that the same reasoning will hold true in a related but different situation. This mistake can happen at a very simple level or at a more complex one. At a simple level, the most common conclusion is that you know you have to reject the reasoning, although it may be difficult to say why. At the more complex level, you may conclude that the reasoning must be accepted even when the results seem to contradict your notion of how the real world works.

No one is quite sure why reasoning and mathematics so often seem to explain the real world. Experience has shown, however, that when the results of reasoning and mathematics conflict with experience in the real world, there is probably a fallacy of some sort involved. As long as you do not know what the fallacy is, the situation is a *paradox*. In some cases, as you will see, the paradox is entirely within mathematics. In others, it is in language or in events in the real world (with associated reasoning). For most paradoxes that are within mathematics, elimination of the fallacious reasoning produces a "purified" mathematics that is a better description of the real world than the "impure" mathematics was. So be it.

In looking at wrong thinking about easy ideas, you will find some cases in which reasoning about the real world is wrong because of a lack of experience with parts of the real world. In other cases, reasoning about mathematics is wrong because certain operations must be ruled out of mathematics. There are also situations in which reasoning is not the right approach to take to the real world.

SEEING IS BELIEVING

You think that the world you see before you is reasonable. Reason can play peculiar tricks on you, though. If you use the right reasons, you may get the wrong answers, at least from the point of view of common sense (or of your uncommon senses). Consider the following argument.

Which has the longer outside rim, a dime or a half-dollar? Going by the judgment of the eye and common sense alone, most persons would answer that the rim of the half-dollar is considerably longer than that of a dime. There is a reason, however, to believe that these lengths are the same.

Probably you have been taught that there is a relation between the distance across a circle and the distance around it. This relation is usually expressed as

$$C = \pi d$$

where C represents the number of units in the circumference, or distance around the circle; d represents the number of units in the diameter, or distance across the circle; and the Greek letter pi, π, stands for a mysterious number slightly larger than 3. How did mathematicians come to this strange conclusion? After all, it is very hard to measure a curve, such as a circle, with any accuracy.

If this theory about the connection between the distance across a circle and the distance around it is true, then the length of the rim of a half-dollar is actually greater than the circumference of a dime, for the diameter of a half-dollar, which is easy to measure because it is straight, is longer than the diameter of a dime. You should not, however, accept a theory just because it was told to you in school. A simple experiment can settle the matter. In fact, if you just visualize the following experiment, you can probably see the answer to the question.

Suppose you bore a small hole through the center of each coin and then put an axle through the holes. The coins should touch, as shown in Fig. 1-1.

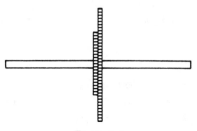

Figure 1-1.

Furthermore, fix the two together so that one will not slip when the other is turned. As you roll the larger coin along a stretch of level tabletop, the small one also turns. If you mark a point on the rim of the half-dollar and, just above that point, another on the rim of the dime, you can keep track of how far you have rolled the coins. Mark the point on the half-dollar *A* and the point above it *B*.

Start with the point *A* touching the table and roll the half-dollar along until *A* touches the table again. Naturally, the distance between the first and second places where *A* touched the table is equivalent to the distance around the rim, if you were careful not to let the half-dollar slip. Since one rotation of the dime would bring *B* around to the starting point again, measuring off the circumference of the dime, you can find whether or not the dime's circumference is greater than, equal to, or less than the length of the rim of the half-dollar. You can see the result in Fig. 1-2 without actually performing the experiment.

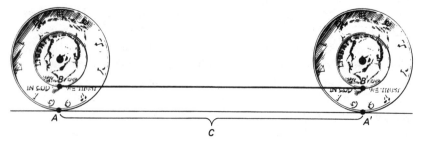

Figure 1-2.

There is no question that the length labeled C is the circumference of the half-dollar. The dime has also rotated once only, and it laid off its circumference, which is also obviously equal to C. Contrary to what you were taught in school, it seems that two circles with different diameters have the same circumference. In fact, it is easy to see that the same method could be used to prove any pair of circles have the same circumference; hence, all circles are the same distance around.

This startling discovery was made by Aristotle more than 2,000 years ago. He did not believe it any more than you do. There must be something wrong with this reasoning (and, indeed, there is).

For one thing, if the experiment is repeated along the edge of the table with the rolling done by the dime, the length of the "circumference of all circles" turns out to be different, as in Fig. 1-3.

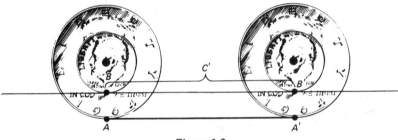

Figure 1-3.

If you have lost faith in rolling coins as a means of measuring the circumference of circles (which would not be surprising), do not bother to take the dime and half-dollar apart to test them along the same stretch of tabletop. Try wrapping a piece of string around the rim of each coin. When the string is unwrapped, it should equal the circumference. If you are quite careful, this method will yield a measure for the circumference that is just a little more than three times the diameter of each coin. Possibly the familiar formula is right after all.

Aristotle's proof, just given, is a classic example of a *paradox*. There are two conclusions. One can be obtained by direct measurement (with a string). The other involves reasoning. Both conclusions seem to be based upon fairly sound evidence. Most persons certainly have a hard time explaining why rolling the two coins with their centers fastened together does not provide the true answer. Although both conclusions are reasonable, they are in direct opposition.

I. The length of the circumference depends upon the diameter.
II. The length of the circumference does not depend upon the diameter.

Since it is reasonably certain that II is false, it is a *fallacy*. When one conclusion of a paradox is false, that conclusion is a fallacy. You cannot be certain that it is a fallacy, however, until you find what mistake was made in the reasoning process.

CAN YOU FIND THE FLAW?

Hint: How do points A and B move as the coins roll along?

Exactly what are you measuring by the method of rolling coins? It appears that you are measuring circumferences, but this surely cannot be. If you were correctly measuring the distances around the coins, the result would not be a fallacy.

When you roll the half-dollar, the point *A* traces a peculiar curve before it comes to rest again on the table, as shown in Fig. 1-4.

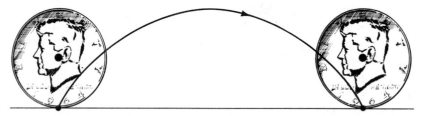

Figure 1-4.

This curve is called the *cycloid*. It has a number of interesting properties. For one thing, the area of a figure bounded by the cycloid and the line along which the coin rolls is exactly three times the area of the disk of the coin. Also, the cycloid in Fig. 1-5 is the curve along which an object would slide in the least time from one point to another point not directly below it.

Figure 1-5.

(You may have thought that a straight line between S and F would be the fastest. In this case, common sense is in error. You can see, however, that sliding down a cycloid would produce the fastest acceleration at the beginning of the trip, thus raising the average velocity. Perhaps some enterprising toy manufacturer will someday make a slide for children in the shape of a cycloid.)

Since the straight line segment AA' between the starting and ending positions of the rolling coin is the shortest distance between the points, the cycloid must be longer. As the coin rotates, then, point A must be moving faster than the imaginary "point" where the coin touches the table. "Point" is in quotation marks because it is a different point (without quotation marks) at every moment of the coin's trip. Yet you can perceive this succession of points as a continuously existing entity—a "point."

It may seem odd, but every point on the rim has a different speed than the "point" touching the table. This strange phenomenon can be accomplished because the speed at which point A moves forward is constantly varying. When A begins its trip on the back side of the coin, it is moving slower than the "point" touching the table; when it reaches the top of the coin, A is impelled by the rotation to move faster than the "point" on the table, and, after it has reached the last quarter of its journey, it begins to slow to the speed of the "point" on the table.

These differences are imparted by the rotation of the coin. You may think of A as a passenger in a bus. As the bus goes forward at a constant speed, A walks to the back (going slower than the bus), remembers to walk to the front to pay his fare (traveling faster than the bus), and walks back again (slower than the bus once more). Despite these complications, A and the bus reach the stop together.

In the connected pair of coins, point B also follows a peculiar path which is another kind of cycloid, as shown in Fig. 1-6.

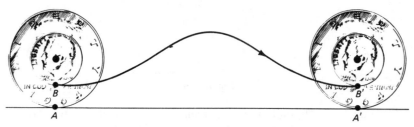

Figure 1-6.

The point B, like the point A, travels at different rates of speed during different parts of the journey. As you can see, the curve has apparently been pulled out of shape by something (in relation to the ordinary cycloid). That something is the forward motion of the coin as a whole. Notice particularly near the ends of the curve how the path has been flattened. Since B is moving slowest in this region, its motion is mainly created from the point being dragged along.

This is the solution, then, or the key to it. Although you are careful not to let the half-dollar slip on the tabletop, the "point" tracing the line segment at the foot of the dime is both rotating and slipping all the time. It is slipping with respect to the tabletop. Since the dime does not touch the tabletop, you do not notice the slipping. If you can roll the half-dollar along the table and at the same time roll the dime (or better yet the axle) along a block of wood, you can actually observe the slipping. If you have ever parked too close to the curb, you have noticed the screech made by your hubcap as it slips (and rolls) on the curb while your tire merely rolls on the pavement. The smaller the small circle relative to the large circle, the more the small one slips. Of course the center of the two circles does not rotate at all, so it slides the whole way. The center is the only point that travels the true circumference of the larger circle. To measure the circumference without taking rotation effects into account, you must measure from the point at the center of the circles before you start to roll the coins to where that point is when you stop. Since the "point" below the center on the tabletop must move at the same speed as the center, however, you are safe in using that "point" in marking off a line segment the length of a circumference. You cannot use the "point" below the center on the small circle to measure the circumference of the small circle, for that "point" also always travels the same distance as the center of both circles, no matter the size of the circle.

In considering the motion of a disk rolling down a hill, a scientist must take the different rates of speed of the various points into account. The point A moves in one continuously varying way, the point B in another, and the disk itself, considered as a whole, in still another. The problem is much more difficult than the consideration of an object sliding down a hill. The physicist uses the various relations discussed with regard to the circle paradox in studying the rolling disk.

When the dime and half-dollar are put together as suggested and rolled along the rim of the dime, the point A on the rim of the half-dollar must move much faster in relation to the "point" beneath the center. This is because it travels farther. Since the rotation speed is constant, point A is sometimes moving backward faster than it is going forward. The path is looped as a result of these short excursions into backward travel.

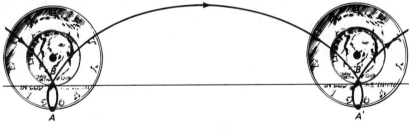

Figure 1-7.

The same general situation occurs also with the rims of the flanges on a railroad train's wheel. The inner part rolls on the track, so the points on the rim describe loops as shown in Fig. 1-7. Even in the fastest train, there are always some points on the train that are moving backward.

The Aristotle circle paradox is connected to another paradox: two circles with the same diameter exist such that the circumference of one is twice that of the other. It is best appreciated if you perform it yourself as a parlor trick.

Take two Lincoln pennies. Place them flat on a table beside each other with the Lincoln heads up, as illustrated in Fig. 1-8.

Figure 1-8.

What do you think will happen if you carefully roll the penny on the left along the half-circumference of the penny on the right, being careful not to let either penny slip? Try to predict the answer without actually performing the operation. Since the penny is rolling along half its circumference, Lincoln ought to turn upside down. (After all, this is what happens if you roll a penny along half its circumference on a tabletop.) Perhaps your experience with the Aristotle paradox will make you suspicious of this.

When the experiment is carefully performed, it appears that Lincoln is right side up at the end of a half turn. Sometimes a person viewing this experiment for the first time will try it over and over, insisting that the coins must be slipping.

One proposed explanation of this trick is that two circles may have the same diameter and different circumferences. The full circumference on one equals half the circumference of the other. Once again, the idea that the ratio of the diameter and circumference is constant is called in question.

CAN YOU FIND THE FLAW?
Hint: What is the path of the center of the moving coin?

As in the previous case, the act of rolling a coin is complicated by the fact that there are two sorts of motion involved: motion along the path of the *whole* coin and rotation around the center. The only point of the coin that moves without being affected by rotation is the center of the circle. In the previous problem, the center moved along a line. This time, the center moves along a curve, as shown in Fig. 1-9.

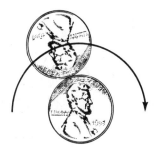

Figure 1-9.

It is not a complicated or unfamiliar curve; it is part of a circle. You can easily calculate the length of this semicircle if you return to the theory that the number of units in the circumference is just π times the number of units in the diameter.

To simplify your computation, measure in "penny-widths" instead of in inches or centimeters. Then the circumference of each one-cent piece is just π penny-widths. The diameter of the circle along which the center of the rolling coin moves is two penny-widths, since half of it consists of one-half the diameter of the stationary penny (its radius) plus one-half the diameter of the rolling coin. That is to say, the radius of the path of the center of the rolling coin equals the diameter of the stationary coin, so the diameter of the path of the center of the rolling coin is twice the diameter of the stationary one.

According to the theory, the number of units in the circumference of the entire circle with a diameter of two penny-widths is just 2π penny-widths. Therefore the distance around the semicircle is π penny-widths. Since the path of the center is the same as the path of the rolling coin, you find that the coin has actually rolled the length of its circumference. It should not be surprising, then, that Lincoln comes to rest right side up.

What would happen if you arranged three pennies in a row with the Lincoln heads right side up and then rolled the outside one along in such a way that the middle one turned freely (Fig. 1-10). The outside coin should roll in a large semicircle. Can you use your knowledge of circles to predict where the head will be when the coins are in a horizontal row again? Work it out in your mind and then try it. If you are careful, you won't be surprised.

Figure 1-10.

You may feel that each of the above examples are cheats, in that you studied mathematical ideas from geometry by manipulating physical objects. You may be saying to yourself, "Sure there are paradoxes and fallacies when you let me make unconscious assumptions about moving coins. But in real mathematics, algebra and that sort of thing, this couldn't happen. Isn't mathematics the one subject where everything works out all right if you follow the rules?" If it is, dear reader, it is only because the operations and ideas that did not work out right were immediately banned by mathematicians. Some operations that seem perfectly reasonable on the surface have caused a great deal of trouble.

IMPERMISSIBLE ACTS

Here is a paradox that has been credited to Augustus De Morgan (1806–1871), but may have originated before his time. De Morgan liked to collect mathematical oddities—or oddities of any kind. De Morgan's collection of odd stories, with a special emphasis on people who had proofs of impossible operations such as squaring the circle with compass and straightedge, was edited after his death to become *A Budget of Paradoxes*. De Morgan was a bit of a paradox himself. He was noted for his eccentricities, but at the same time was a gifted algebraist and teacher. He also had a major role in developing mathematical logic. Clearly, anyone that logical knew exactly what was wrong with the following argument, even though it is attributed to him.

Let $x = 1$. Then x is also equal to 0.

$$x = 1$$

Multiply both sides by x $\qquad x^2 = x$

Subtract 1 from each side $\qquad x^2 - 1 = x - 1$

Divide each side by $x - 1$

Since this is harder, it may help to show the operation in full for those who have forgotten their algebra.

$$\frac{x^2 - 1}{x - 1} = \frac{x - 1}{x - 1}$$

Factor the numerator on the left side. Then some factors cancel out.

$$\frac{(x + 1)\,(\cancel{x - 1})}{\cancel{x - 1}} = \frac{\cancel{x - 1}}{\cancel{x - 1}}$$

$$x + 1 = 1$$

Subtract 1 from each side $\qquad x = 0$

```
CAN YOU FIND THE FLAW?
Hint: Which step was "illegal?"
```

Since in this case you are dealing with abstractions of algebra, there is no chance that your senses were fooled, as they may have been in looking at coins being rolled. Clearly, x cannot equal both 1 and 0 at the same time. In fact, there are many kinds of equations that are true when x equals either 1 or 0—for instance, $x\,(x-1) = 0$—but this is nothing like that situation. It is specified that $x = 1$. Then a logical argument is offered that $x = 0$. Therefore, the conclusion $x = 0$ and reasoning that led to $x = 0$ constitute a fallacy.

Actually, the argument is more algebraic than logical and the flaw in the argument is the hidden use of an "illegal" step. You may recall that it is not permissible to divide by zero. In fact, division by zero is ruled out because it leads to paradoxical situations. Division is defined as the opposite (more properly, the *inverse*) of multiplication. Hence, $a \div b = c$ means $c \times b = a$. Suppose $b = 0$. If a is *not* 0 also, then $c \times b = a$ has no solution. What is worse is that if $a = 0$, then c could be any number.

Look at an example with specific numbers. If $5 \div 0 = c$, then $c \times 0 = 5$. But every number times zero is equal to zero, so there is no number c such that $c \times 0 = 5$. On the other hand, if $0 \div 0 = c$, then $c \times 0 = 0$. This is true of any number c.

At some time, you may have learned that any number divided by zero is infinity, symbolized by ∞. If so, you were misinformed. In some cases, of course, the value of some expression A/B may increase without bound (which is what ∞ actually means) as B gets closer to zero. This is discussed in more detail in the next chapter. In any case, the De Morgan paradox does not involve that situation.

Look at the calculations again, but with x replaced by 1 in each step.

$$1 = 1$$

Multiply both sides by 1 $\qquad 1^2 = 1$

or $\qquad\qquad\qquad\qquad\qquad 1 = 1$

Subtract 1 from each side $\qquad 1 - 1 = 1 - 1$

or $\qquad\qquad\qquad\qquad\qquad 0 = 0$

Divide each side by $1 - 1$ $\qquad \dfrac{0}{1-1} = \dfrac{0}{1-1}$

or $\qquad\qquad\qquad\qquad\qquad \dfrac{0}{0} = \dfrac{0}{0}$

As noted above, $0/0$ (which is the same as $0 \div 0$), if allowed in mathematics, could mean any number whatsoever.

Furthermore, you can generalize the fallacy to show that any number, a, is equal to zero. If $x = a$, then

Multiply by x	$x^2 = ax$
Subtract a^2 from each side	$x^2 - a^2 = ax - a^2$
Divide each side by $x - a$	$x + a = a$
Subtract a from each side	$x = 0$

It should be clear from the version with numbers substituted for letters that the illegal step is the division by $x - a$ ($= 0$). Before division by zero, all the results were true. Afterwards, they could mean anything.

Another look at this fallacy, however, shows that defining $0/0$ as always equal to zero might solve the problem equally as well as ruling out division by zero. For example, if $0 \div 0 = c$ means $c \times 0 = 0$, then making $0/0$ equal to zero would satisfy both statements. Also, if you obtained 0 instantly in the first example when you divided by $x - 1$ (or by $x - a$ in the second example), then the problem would be over. You would have $0 = 0$, no matter how it was expressed. The rest of the manipulation would be just changing the form of $0 = 0$.

There are a number of ways, however, to show that this does not work. The easiest is to go back to the original De Morgan paradox. Let $x = 1$.

Multiply by x	$x^2 = x$
Subtract 1	$x^2 - 1 = x - 1$
Divide by $x - 1$ (now you have	$\dfrac{x^2 - 1}{x - 1} = \dfrac{x - 1}{x - 1}$
$0 = 0$ if $0/0 = 0$ is true)	$x + 1 = 1$
But $x = 1$, so	$2 = 1$

This is a result that cannot legitimately be obtained from $0/0 = 0$. Or, you could continue this process further before you substituted 1 for x:

	$x + 1 = 1$
Multiply each side by 2	$2x + 2 = 2$
Substitute 1 for x	$2 + 2 = 2$
or	$4 = 2$

There is a rule of logic that states that any conclusion whatsoever can be obtained from a false statement. Here you see an instance of it. The statement could be written

$$x = 1 \quad and \quad x + 1 = 1$$

The use of *and* means that both $x = 1$ *and* $x + 1 = 1$ must be true at the same time. This is not possible if mathematics is to be meaningful, for reasoning from that false statement can produce $1 = 0$, $2 = 1$, $4 = 2$, or $a = b$ for any numbers a and b.

Here is one way to get from $x = 1$ and $x + 1 = 1$ to $a = b$. If $x + 1$ is raised to a suitable power and 1 substituted for x, the power can be made to have any value you desire. For example, $(x + 1)^2 = x^2 + 2x + 1$. When 1 is substituted for x, the result is $1 + 2 + 1$, or 4. Similarly, $(x + 1)^3 = x^3 + 3x^2 + 3x + 1$, so when 1 is substituted, the expression is equal to $1 + 3 + 3 + 1$, or 8. This should not be a surprise, for in each case you get as your result the corresponding power of 2, for $x + 1$ is really $1 + 1$ in disguise. Use of a rational exponent, such as $1/2$, gives other values. In fact, $2^{1/2}$ is a number near to 1.414. Now choose some exponent n so that $(x + 1)^n$ is equal to a/b when 1 is substituted for x.

	$x + 1 = 1$
Raise to the nth power	$(x + 1)^n = 1^n$
or	$(x + 1)^n = 1$
Substitute 1 for x	$\dfrac{a}{b} = 1$
or	$a = b$

Now consider a different impermissible act. Although division by zero is never permissible, there are some operations in algebra that are permissible *almost all* of the time, but which are not permissible in specific instances. For example, the equation

$$\frac{3}{x} = \frac{3}{2}$$

can readily be solved by noting that if

$$\frac{a}{b} = \frac{a}{c}$$

then

$$b = c$$

So the solution to $3/x = 3/2$ is $x = 2$.

Furthermore, you can solve the equation

$$\frac{x - 3}{x - 1} = \frac{x - 3}{x - 2}$$

by noting that the solution is the same as the solution to

$$x - 1 = x - 2$$

But this is the same as saying $-1 = -2$ or $1 = 2$.

CAN YOU FIND THE FLAW?
(No hint; this one is too easy.)

One way to tackle this problem is to solve it by conventional means.

$$\frac{x - 3}{x - 1} = \frac{x - 3}{x - 2}$$

Multiply each side of the equation by $(x - 1)(x - 2)$.

$$\frac{(x - 1)(x - 2)(x - 3)}{x - 1} = \frac{(x - 1)(x - 2)(x - 3)}{x - 2}$$

Cancel the common factors in the numerators and denominators.

$$(x - 2)(x - 3) = (x - 1)(x - 3)$$

or $\qquad x^2 - 5x + 6 = x^2 - 4x + 3$

Subtract x^2 $\qquad -5x + 6 = -4x + 3$

or $\qquad\qquad\quad 3 = x$

Now the error that led to the fallacy should be apparent. Since x is 3, the original equation represented

$$\frac{3 - 3}{3 - 1} = \frac{3 - 3}{3 - 2}$$

or

$$\frac{0}{2} = \frac{0}{1}$$

This is perfectly reasonable since there is no rule that prohibits dividing into zero (just division *by* zero). However,

$$\frac{0}{2} = \frac{0}{1}$$

does not mean that $2 = 1$ is true. The rule that if

$$\frac{a}{b} = \frac{a}{c}$$

then $b = c$ must be modified to say *if a is not zero*.

Here is a very closely related situation, but with a significant difference. You know that

$$\frac{a}{b} = \frac{c}{d}$$

implies

$$\frac{b}{a} = \frac{d}{c}$$

Suppose you use this rule to change

$$\frac{x - 3}{x - 1} = \frac{x - 3}{x - 2}$$

into

$$\frac{x - 1}{x - 3} = \frac{x - 2}{x - 3}$$

This equation can be approached from the naive point of view by thinking, "The denominators are equal, so the numerators must also be equal." Once again, you obtain

$$x - 1 = x - 2$$

or

$$1 = 2$$

On the other hand, you can proceed to solve it by multiplying both sides by $x - 3$, a rule of algebra that seems time tested: *if you multiply both sides of an equation by the same number, the result is an equation with the same solution.*

Multiply each side by $x - 3$

$$\frac{(x - 3)(x - 1)}{x - 3} = \frac{(x - 3)(x - 2)}{x - 3}$$

Canceling gives $\qquad x - 1 = x - 2$

or $\qquad 1 = 2$

That approach does not get you very far either.

The truth is that the equation

$$\frac{x - 3}{x - 1} = \frac{x - 3}{x - 2}$$

has a perfectly good solution, but the similar equation

$$\frac{x - 1}{x - 3} = \frac{x - 2}{x - 3}$$

has *no* solution. If it had a solution, the solution would be $x = 3$. However, substituting 3 for x gives

$$\frac{3 - 1}{3 - 3} = \frac{3 - 2}{3 - 3}$$

or

$$\frac{2}{0} = \frac{1}{0}$$

which is not a permissible operation.

These problems can be generalized to note that

$$\frac{x - a}{x - b} = \frac{x - a}{x - c}$$

with b not equal to c always has the solution

$$x = a$$

while

$$\frac{x - b}{x - a} = \frac{x - c}{x - a}$$

with b not equal to c never has a solution. (In case b is equal to c, what happens?)

Look at one more equation of a related type before leaving this class of paradoxes.

$$\frac{x - 3}{x - 3} = \frac{x - 1}{x - 2}$$

Solving it one way, note that

$$\frac{x - 3}{x - 3} = 1$$

so the equation can be rewritten as

$$1 = \frac{x - 1}{x - 2}$$

or

$$x - 2 = x - 1$$
$$1 = 2$$

But suppose that you don't notice that $(x - 3)/(x - 3) = 1$. Instead, you decide to multiply both sides by $(x - 3)(x - 2)$. You get

$$\frac{(x - 3)(x - 2)(x - 3)}{x - 3} = \frac{(x - 3)(x - 2)(x - 1)}{x - 2}$$

Canceling out the common factors produces

$$(x - 3)(x - 2) = (x - 3)(x - 1)$$

Completing the multiplication gives

$$x^2 - 5x + 6 = x^2 - 4x + 3$$

You solved this equation earlier. The solution is

$$3 = x$$

Therefore, $x - 3$ would have to equal zero, so the equation has no solution. Even though the algebraic procedure produces a solution, the solution must be rejected. In fact, algebraic procedures always are limited to producing "candidates" for solution. The "candidate" may not match the conditions of the problem being solved, or it may be a false solution that must be rejected for other reasons.

Look again at

$$(x - 3)(x - 2) = (x - 3)(x - 1)$$

Since $x - 3$ is a common factor on both sides, it would appear that the obvious step would be to solve the equation by dividing by $x - 3$. But that produces

$$x - 2 = x - 1$$

or

$$1 = 2$$

Nevertheless, the equation

$$(x - 3)(x - 2) = (x - 3)(x - 1)$$

has a perfectly good solution, namely $x = 3$. Substituting 3 for x gives

$$(3 - 3)(3 - 2) = (3 - 3)(3 - 1)$$

or

$$0 \times 1 = 0 \times 2$$

which is certainly true. Dividing the equation by $x - 3$ is a way of dividing by zero, so the result is meaningless.

Note that the original statement of the rule for changing from one equation to another should have read: *if you multiply or divide both sides of an equation by the same* nonzero *number, the result will be an equation with the same solution.* If the number is zero, however, the resulting equation may not have the same solution or it may have extra solutions. While multiplication by zero does not have quite the devastating effect of dividing by zero, it often changes the solution.

In fact, for any equation

$$A = B$$

where A and B are expressions involving numbers and the variable x, the equation

$$(x - a)A = (x - a)B$$

where a is some number, has the solution $x = a$ (because that makes the equation $0 = 0$, which is true). Unless a is a solution of $A = B$ in the first place, multiplication by $x - a$ changes the solution. If the original equation is always false, say

$$1 = 2$$

the equation after multiplication has a solution. Say you multiply $1 = 2$ by $x - 1$.

$$x - 1 = 2(x - 1)$$
$$x - 1 = 2x - 2$$
$$1 = x$$

If the original equation had a solution, say $x = 2$, then the equation after multiplication by $x - 1$ would have two solutions: 1 and 2. You should be aware of this when you start playing around with algebra.

Here is another proof that any two numbers a and b are equal. This one is based on taking the square root of both sides of an equation. The operation is based on the notion that if both sides of an equation represent the same number, then the square roots also represent the same number.

Consider any two different numbers a and b. Call them sum $2c$.

$$a + b = 2c$$

Multiply each side by $a - b$

$$(a - b)(a + b) = (a - b) \cdot 2c$$

or

$$a^2 - b^2 = 2ac - 2bc$$

Add $b^2 + c^2 - 2ac$ to each side of the equation.

$$a^2 - 2ac + c^2 = b^2 - 2bc + c^2$$

Each side of the equation is now in the form of a perfect square.

$$(a - c)^2 = (b - c)^2$$

Take the square root of each side.

$$a - c = b - c$$

or

$$a = b$$

> **CAN YOU FIND THE FLAW?**
> **Hint:** There are two of them.

After reading the discussion just prior to this fallacy, it should be easy to spot the first flaw. When you multiply by $a - b$, you introduce the solution $a = b$ into the equation. Notice that

$$a^2 - b^2 = 2ac - 2bc$$

is true when a and b are equal ($0 = 0$). There should be another solution, however. From the original equation $a + b = 2c$, you know that $a = 2c - b$. What happens to that solution? Here is where the second flaw in the argument comes into the picture.

When you take the square root of a number such as 25, you normally think of the answer as 5. And, in fact,

$$\sqrt{25} = 5$$

is true. There is another square root, however: -5, for $(-5)(-5) = 25$. By definition it is not true that

$$\sqrt{25} = -5$$

for the sign $\sqrt{}$ is defined to be the positive square root. This definition of $\sqrt{}$ is useful, but it reinforces the habit of thinking only of the positive square root.

Thus, when

$$(a - c)^2 = (b - c)^2$$

is changed to

$$a - c = b - c$$

only the positive square roots appear. However, the equation has solutions for all four combinations of positive and negative square roots.

$a - c = b - c$	(both "positive")
$-(a - c) = -(b - c)$	(both "negative")
$-(a - c) = b - c$	
$a - c = -(b - c)$	

The last two equations mix "positive" and "negative." The words *positive* and *negative* are put in quotation marks because you don't know which variables *really* represent positive or negative numbers until you know the values of a, b, and c. For example, $-x$ represents a negative number, if x is 2; but $-x$ represents a positive number if x is -2. So $-x$ is not really negative, but for this purpose can be called "negative."

If you solve the four equations, the results reveal what happened to the missing solution.

$a - c = b - c$	has the solution	$a = b$
$-(a - c) = -(b - c)$	has the solution	$a = b$
$-(a - c) = b - c$	has the solution	$a = 2c - b$
$a - c = -(b - c)$	has the solution	$a = 2c - b$

This all seems simple enough when you think it through carefully today, but imagine what such a result would have meant to algebraists who worked before the invention of negative numbers. If you do not know about negative numbers, then it is very difficult, maybe impossible, to explain what happened to the original solution, or $a = 2c - b$. This is an instance of how a paradoxical result can be handled by recognizing it as a fallacy. In other words, what would have been a strange paradox to a mathematician of any time before the age of Brahmagupta, who put together a set of rules for dealing with negative numbers during the seventh century A.D., seems more like an error today.

In fact, some of the earlier fallacies in this chapter would have been paradoxes for Brahmagupta, for he believed that $0/0 = 0$, as was suggested earlier as a possibility.

Don't blame Brahmagupta for making mistakes about zero. The idea was just coming into use around his time. The oldest inscription showing the cardinal number 0 for zero dates from A.D. 876, after Brahmagupta's death.

How would mathematicians explain the results of this section without a notion of 0? I think that they could not. Before 0 was in-invented, these were genuine paradoxes.

Here is a fallacy that does not involve division by zero. You have been using the relationship

$$(x + 1)^2 = x^2 + 2x + 1$$

which is true for any x. By playing with this statement that is true for all x, you can get a statement that is false for all x. First rewrite the equation as

$$(x + 1)^2 - (2x + 1) = x^2$$

Subtract $x(2x + 1)$ from each side of the equation.

$$(x + 1)^2 - (2x + 1) - x(2x + 1) = x^2 - x(2x + 1)$$

On the left side, you can factor out $(2x + 1)$ from the last two terms.

$$(x + 1)^2 - (2x + 1)(x + 1) = x^2 - x(2x + 1)$$

Now add $1/4(2x + 1)^2$ to each side of the equation.

$$(x + 1)^2 - (2x + 1)(x + 1) + \frac{1}{4}(2x + 1)^2 = x^2 - x(2x + 1)$$
$$+ \frac{1}{4}(2x + 1)^2$$

The next step is conceptually easy, but the algebraic manipulation is hard to follow. You know that $(a - b)^2 = a^2 - 2ab + b^2$. On the left side of the equation, let

$$x + 1 = a$$

and

$$\frac{1}{2}(2x + 1) = b$$

Then

$$(2x + 1)(x + 1) = 2 \cdot \frac{1}{2}(2x + 1)(x + 1) = 2ba = 2ab$$

and

$$\frac{1}{4}(2x + 1)^2 = b^2$$

So

$$(x + 1)^2 - (2x + 1)(x + 1) + \frac{1}{4}(2x + 1)^2 = \left[(x + 1) - \frac{1}{2}(2x + 1)\right]^2$$

The same thing can be done with the right side. The resulting equation (left side equal to right side) is

$$\left[(x + 1) - \frac{1}{2}(2x + 1)\right]^2 = \left[x - \frac{1}{2}(2x + 1)\right]^2$$

Now take the square root of each side.

$$(x + 1) - \frac{1}{2}(2x + 1) = x - \frac{1}{2}(2x + 1)$$

or

$$x + 1 = x$$

CAN YOU FIND THE FLAW?
Hint: It's one you've seen before.

It is instructive with any paradox of this type to choose some real number (instead of a variable) and follow the same steps. For simplicity in computation, you can choose 0 this time (although 0 and 1 are often too special to use in checking problems). Then

$$(x + 1)^2 = x^2 + 2x + 1$$

becomes

$$(0 + 1)^2 = 0^2 + 2 \cdot 0 + 1$$

or

$$1^2 = 1$$

which is certainly true. You rewrite this as $1^2 - 1 = 0$. Now you subtract $0(2 \cdot 0 + 1)$, or 0, from each side. No change. Factoring also brings no change. When you add $1/4(2x + 1)^2$ to each side, the equation changes from $0 = 0$ to $1/4 = 1/4$. Then that result is written as the equality of two squares.

$$\left[(x + 1) - \frac{1}{2}(2x + 1)\right]^2 = \left[x - \frac{1}{2}(2x + 1)\right]^2$$

When 0 is substituted, this becomes

$$\left[(0 + 1) - \frac{1}{2}(2 \cdot 0 + 1)\right]^2 = \left[0 - \frac{1}{2}(2 \cdot 0 + 1)\right]^2$$

or

$$\left(1 - \frac{1}{2}\right)^2 = \left(0 - \frac{1}{2}\right)^2$$

$$\left(\frac{1}{2}\right)^2 = \left(-\frac{1}{2}\right)^2$$

This is certainly true, but the equation found by taking the square root, $1/2 = -1/2$, is not.

Therefore, you get the false statement

$$x + 1 = x$$

or

$$0 + 1 = 0$$

As in the earlier example, taking the square root of both sides of an equation should produce four separate equations:
If

$$x^2 = y^2$$

then it can be said that at least one of the following equations is true

$$x = y$$
$$-x = -y$$
$$-x = y$$
$$x = -y$$

Two of these are equivalent, so this can be reduced to

$$x = y$$
$$x = -y$$

Unless $x = 0$, either one of these two equations is false or there are two solutions. The only way of telling which solution is false and which is true is to substitute the two solutions into the original equation—before the operation of taking the square root is performed.

Therefore, in the derivation of this fallacy (it was a paradox when you did not know what had gone wrong because you had forgot the rule about taking the square root or were unfamiliar with negative numbers), the last step should have been written as follows:

Now take the square root of each side.

$$(x + 1) - \frac{1}{2}(2x + 1) = x - \frac{1}{2}(2x + 1)$$

or

$$(x + 1) - \frac{1}{2}(2x + 1) = -x + \frac{1}{2}(2x + 1)$$

but not necessarily both.

Now the first equation on p. 29 is equivalent to $x + 1 = x$, which is certainly false, so the second must be true. The second equation is equivalent to

$$(x + 1) - (2x + 1) = -x$$

or

$$-x = -x$$

No problem.

ELIMINATING PARADOX BY DEFINITION

At first thought, you might believe from the definition of x^2 that $\sqrt{x^2} = x$. However, in the discussion of how to prove $a = b$ if $a + b = 2c$, you were reminded that by definition $\sqrt{}$ is used to represent only the positive square root of a number. Thus, if x is -3, then $\sqrt{x^2}$ is not -3, but $+3$. This seems odd (paradoxical?) on first encounter, but it is easy to get used to. The absolute value function is used in careful work in mathematics to avoid error in this situation. The absolute value of a number x is the same as x if x represents a positive number. If x represents a negative number, however, the absolute value of x is $-x$, which is a positive number. The symbol $| \ |$ is used for absolute value, so

$$|+3| = 3$$
$$|-3| = 3$$
$$|0| = 0$$

The last equation needs to be defined separately, as zero is neither positive nor negative. With the absolute value sign, you can write

$$\sqrt{x^2} = |x|$$

and be sure that you are safe. Most of the time.

Consider the imaginary numbers. You may recall that complex and imaginary numbers are needed to solve many equations. The simplest such equation is $x^2 + 1 = 0$. It is solved by inventing a number, often symbolized by $\sqrt{-1}$, as one solution and the related number $-\sqrt{-1}$ as the other solution. (In general, equations with x^2 as the highest power of x have two solutions.) The number $\sqrt{-1}$ is defined by the equation

$$(\sqrt{-1})^2 = -1$$

Now consider the following. The square root of a product is equal to the product of the square roots. For example,

$$\sqrt{4} \cdot \sqrt{9} = \sqrt{4 \cdot 9}$$
$$2 \cdot 3 = \sqrt{36}$$
$$6 = 6$$

More generally, $\sqrt{x} \cdot \sqrt{y} = \sqrt{xy}$. If this were applied to $\sqrt{-1}$, it would mean that

$$(\sqrt{-1})^2 = \sqrt{-1} \cdot \sqrt{-1} = \sqrt{(-1)(-1)}$$
$$= \sqrt{1}$$
$$= 1$$

Note that even though -1 is a square root of 1, $\sqrt{1}$ is defined to be 1, not -1. But $(\sqrt{-1})^2$ is defined to be -1. Therefore,

$$1 = -1$$

CAN YOU FIND THE FLAW?

Hint: Look at the title of this section.

When the definition gets you in trouble, change the definition. As long as you are dealing with positive values for x and y, the rule that $\sqrt{x} \cdot \sqrt{y} = \sqrt{xy}$ presents no problems. In fact, if x is negative and y positive or vice versa, there are no difficulties. But when x and y are both negative, a strange thing happens.

$$\sqrt{-4} \cdot \sqrt{-9} = \sqrt{(-4)(-9)} = \sqrt{36} = 6$$

$$\sqrt{-4} \cdot \sqrt{-9} = \sqrt{-1}\sqrt{4} \cdot \sqrt{-1}\sqrt{9} = (\sqrt{-1})^2\sqrt{36} = (-1)(6) = -6$$

In the second equation, $\sqrt{-4}$ is replaced by $\sqrt{-1} \cdot \sqrt{4}$. Similarly, $\sqrt{-9}$ is replaced by $\sqrt{-1} \cdot \sqrt{9}$. As noted above, no problems arise with the rule $\sqrt{x} \cdot \sqrt{y} = \sqrt{xy}$ when one variable represents a negative number and the other a positive number. And there are no problems when both variables represent positive numbers. Now $(\sqrt{-1})^2$ was *defined* to mean -1 (even though the paradox shows that it could mean 1 also). This definition, as it stands, does not seem to work. Mathematicians have to introduce a whole set of auxiliary definitions to make the system work consistently.

FIRST: Throw out $\sqrt{-1}$. It clearly causes problems. Replace $\sqrt{-1}$ with the letter i in all cases. The solutions to $x^2 + 1 = 0$ are i and $-i$, not $\sqrt{-1}$ and $-\sqrt{-1}$.

SECOND: Define $\sqrt{-|a|}$ as $i\sqrt{|a|}$ in all cases. So, if $\sqrt{-4}$ arises in a problem, it should be replaced immediately by $i\sqrt{4}$.

THIRD: Restrict the rule that $\sqrt{x} \cdot \sqrt{y} = \sqrt{xy}$ to cases where neither x nor y is negative. Also restrict the similar rule for division in the same way.

In this way, the paradoxes are defined out of existence. For example, you cannot write $\sqrt{-1} \cdot \sqrt{-1}$ any longer. You must write $i \cdot i$, which has an unambiguous meaning. Similarly, $\sqrt{-4} \cdot \sqrt{-9}$ *must be* replaced by $i\sqrt{4} \cdot i\sqrt{9}$, so the value of the product is -6 with no questions.

Here is another example of the kind of problems that can arise when these rules are not followed.

$$\sqrt{-1} = \sqrt{-1}$$

Replace each occurrence of -1 with a fraction form of -1.

$$\sqrt{\frac{1}{-1}} = \sqrt{\frac{-1}{1}}$$

Use the division rule that $\sqrt{x/y} = \sqrt{x}/\sqrt{y}$.

$$\frac{\sqrt{1}}{\sqrt{-1}} = \frac{\sqrt{-1}}{\sqrt{1}}$$

This equation is equivalent to the following

$$\sqrt{1} \cdot \sqrt{1} = \sqrt{-1} \cdot \sqrt{-1}$$

Therefore, since $(\sqrt{-1})^2 = -1$,

$$1 = -1$$

Clearly, this unfortunate result could not have occurred if you had begun by replacing $\sqrt{-1}$ with i.

Mathematicians have gone somewhat further, in fact, in setting up a system to make sure that no paradoxes arise in dealing with $\sqrt{-1}$. They define i as the *ordered pair* (0, 1). By defining addition of ordered pairs as

$$(a, b) + (c, d) = (a + b, c + d)$$

and multiplication of ordered pairs as

$$(a, b) \cdot (c, d) = (ac - bd, ad + bc)$$

they get a system, called the *complex numbers*, that can be shown to be as free from contradictions as the real numbers to which you are accustomed. If any problems of interpretation occur, you can always return to the formal definitions based on ordered pairs. For example, instead of worrying about whether $\sqrt{-1} \cdot \sqrt{-1}$ is equal to 1 or to -1, you know for sure that $(0, 1) \cdot (0, 1) = (0 \cdot 0 - 1 \cdot 1, 0 \cdot 1 + 1 \cdot 0) = (-1, 0)$. In the complex numbers, of course, $(-1, 0)$ plays the same part as -1 in the familiar real numbers.

When you assume that the same rules hold for new mathematical entities, such as $\sqrt{-1}$, as hold for familiar ones, trouble can result. Such trouble occurs later in this book in situations that are not so easy to resolve as the problems concerning $\sqrt{-1}$. In fact, there are problems with extensions of mathematical systems that *cannot* be resolved by defining them away. Yet these problems occur with extensions of mathematics that seem more natural than the step from the solutions of $x^2 - 1 = 0$ to the solution of $x^2 + 1 = 0$.

AN UNEXPECTED PARADOX

You cannot always trust physical experiments, as the example of the experiment with circles showed. You cannot always trust mathematics either, for it can mislead you unless you define away the problem areas. However, you surely can trust pure logic—no phony experiments or odd mathematical operations. Let's see.

During World War II the Swedish Broadcasting Company made the following announcement on the radio:

> A CIVIL-DEFENSE EXERCISE WILL BE HELD THIS WEEK. IN ORDER TO MAKE SURE THAT THE CIVIL-DEFENSE UNITS ARE PROPERLY PREPARED, NO ONE WILL KNOW IN ADVANCE ON WHAT DAY THS EXERCISE WILL TAKE PLACE.

A Swedish mathematician, Lennart Ekbom, immediately recognized something odd about this announcement. He discussed the situation with his class at Östermalms College. From there, it spread around the world. By 1948 it had reached print in the British magazine *Mind*. In 1951, Michael Scriven announced, "A new and powerful paradox has come to light." Do you see what the essence of the problem is?

Consider the announcement by the Swedish Broadcasting Company again. Suppose that it was made on Monday morning (no one seems to recall exactly when it was made). Then the exercise must take place sometime before the following Monday. Yet, it cannot take place on Sunday, for by then people will know that it has to take place Sunday. Since it will be within the week and will be unexpected, Sunday is ruled out.

Also, Saturday is ruled out. Since the exercise cannot take place on Sunday, it also cannot happen on Saturday, for you would know it in advance. If it has not happened during the week so far, and if it cannot happen Sunday, then people would know that it was going to take place on Saturday. So it cannot.

Friday is no good either. With both Saturday and Sunday out, if it has not happened by Friday, everyone will expect it. Similar reasoning applies to Thursday, Wednesday, Tuesday, and Monday itself. Therefore, the civil-defense exercise cannot happen at all.

But on Wednesday morning, the air-raid sirens wail and the exercise takes place anyway. Irrefutable logic has been refuted by reality.

> **CAN YOU FIND THE FLAW?**
> **Hint**: What you don't know can't always be found by reason.

There are many versions of this paradox. All are essentially the same. Two people are required. One says that something will happen and that it will be unexpected. The other person reasons that these conditions are contradictory. Therefore, the event cannot happen. But it happens anyway.

One of the clearest versions comes from Martin Gardner, columnist for *Scientific American*. A loving husband tells his wife that she will receive an unexpected gift for her birthday. It will be a gold watch. The husband is the person who has set the conditions.

Now the wife uses logic. Her husband would not lie to her. Since he has said the gift would be unexpected, it will be unexpected. But she now expects a gold watch. Therefore, it cannot be a gold watch.

But, of course, it is.

And it is unexpected, for she had used logic to show that it could not be a gold watch.

There have been numerous discussions of this paradox since it first "came to light." In general, they have concurred with the conclusion that Willard Van Orman Quine, the Harvard logician, reached in 1953. Quine identified four possible conclusions that can be reached in these situations. For the Swedish civil-defense exercise, these are as follows:

I. The exercise will occur this week and be expected.
II. The exercise will not occur this week but it will be expected.
III. The exercise will not occur this week and be unexpected.
IV. The exercise will occur this week but it will be unexpected.

The Swedes are renowned for being truthful, so conclusion I is false because the broadcast said the exercise would be unexpected. Similarly, you can rule out conclusions II and III, for each does not conform to the conditions stated in the announcement. Therefore, only IV can be true.

This analysis is helpful, but not very satisfying. It tells you what you knew to be true anyway, but it does not attack the problem of the logical argument that statement IV is contradictory.

There is an old story about two madmen that is relevant here. One comes to the other in the mental ward with his hands cupped as if he were holding a hidden object (such as an egg). He says, "I'll bet you can't tell what I am holding here." The other madman promptly says, "The New York Philharmonic." Taken aback, the first madman peeks cautiously into his hands and asks, "Who's conducting?"

From the point of view of Logical Thinker A, Logical Thinker B is always a madman. For Logical Thinker B can always know something that is inaccessible to Logical Thinker A, namely, what Logical Thinker B is thinking. This information is beyond any reasoning on A's part. In short, logic does not apply to another person's thoughts. The wife cannot reason about her husband's peculiar statement that the gold watch will be unexpected. This is as forbidden as division by zero.

The story of the two madmen is a joke because it violates our sense that one person can know something unexpressed that is taking place in another's mind. If I come to you with my hands cupped and say that there is an egg inside, you have no way of knowing whether I am telling the truth or not. Therefore, if I open my hands and there *is* an egg, it will be unexpected (even though I told you it was there).

But wait a minute? If you know me to be a truthful person, you would have expected the egg.

And if I had said the New York Philharmonic was in my hands, you would have decided that I had gone bats.

But in none of the cases—egg, no egg, or New York Philharmonic— could you have determined what I was thinking by reasoning. You could rule out the New York Philharmonic by reasoning, but that is a perfectly legitimate use of logic to deal with the real world. You cannot exclude egg or no egg by reasoning.

Trying to determine when the Swedish civil-defense exercise will take place by reasoning is just a more complicated case of trying to determine by reasoning whether I am holding an egg in my hands. The Swedish civil-defense authority knows something about the exercise that the general public does not. I know something about the egg that you do not. Reasoning does not apply.

It may seem odd to include this paradox in a chapter that is largely about mathematics. Whether logic is a branch of mathematics, mathematics a branch of logic, or neither of the above has been debated for years. In this case, it is included here because it is essentially a simple fallacy. When it first appeared, so very recently as compared to the other fallacies in the chapter, it seemed to many people to be a genuine paradox. It was not long, however, until logicians found the flaw.

In Chapter 2, you will encounter some paradoxes that also were degraded to fallacy status, although it took a lot longer. In later chapters, you will encounter some paradoxes that cannot be reevaluated as fallacies. These come in two types: *a*) paradoxes that you have to learn to live with and *b*) paradoxes that you have to learn to avoid. In a sense, however, type *b* paradoxes are more sophisticated versions of the fallacies caused by incorrect operations in mathematics. The only thing to be done with them is to define them away, as you did with division by zero or with $\sqrt{-1}$; then, since these paradoxes are ruled out, use of them in reasoning becomes a fallacy.

This situation is a paradox itself.

2
Thinking Wrong about Infinity

There is no sharply drawn line between those contradictions which occur in the daily work of every mathematician, beginner or master of his craft, as a result of more or less easily detected mistakes, and the major paradoxes which provide food for logical thought for decades and sometimes centuries.

Nicholas Bourbaki

In Chapter 1, it was noted that the commonly held belief that $x/0 = \infty$, where ∞ is the symbol for infinity, is wrong on several counts. Since the remainder of this book deals largely with infinity in one form or another, it will be essential to have a good understanding of what infinity is all about. Infinity is not easy to think about, however, so you will approach it by bits and pieces. For now, the most important thing to understand is that the word *infinity* suffers from the problem that words in general use always have—there are many meanings that are rather poorly defined. To handle the complexities of infinity, it will be necessary to introduce technical terms that have very specific meanings. Otherwise, you might think you were dealing with *infinity*$_1$ when you are actually dealing with *infinity*$_2$ or some other infinity altogether.

THE TROUBLE WITH "AND SO FORTH"

It is always easiest to begin with the natural numbers, which in this book are $1, 2, 3, \ldots$ (some mathematicians include 0 as a natural number, but for the purposes here it is best to say *whole numbers* when 0 is to be included along with $1, 2, 3, \ldots$).

The three dots after 3 are a mathematician's way of saying "and so forth" or "etc." For example,

$$1, 2, 3, \ldots, 10$$

means

$$1, 2, 3, 4, 5, 6, 7, 8, 9, 10$$

Using three dots in this way is not a very rigorous way to say "and so forth." It is a little-known mathematical fact that for any sequence of numbers there is a formula that can be found that will produce the sequence as far as it is stated and then produce some number that is not what you had in mind at all. Thus, the use of sequences on intelligence tests is highly suspect. Suppose, for example, that you are taking an intelligence test and are presented with the sequence 1, 2, 3, with the idea that you should supply the next number. If you answered "4" and I marked you wrong, then you would be upset. However, I was thinking of a sequence whose first four terms are 1, 2, 3, 10. This sequence can be found by replacing n in the formula $(n - 1)(n - 2)(n - 3) + n$ with the numbers 1, 2, 3, and 4 in order.

Try another one. What is the next number in the sequence

$$1, 4, 9, 16, \ldots ?$$

While 25 might come to mind, it is actually 49, for I was thinking of the sequence you get when you replace n in

$$(n - 1)(n - 2)(n - 3)(n - 4) + n^2$$

by 1, 2, 3, 4, and 5 in order. If you wanted 25 as the next number, then you probably wanted to replace n in n^2 by 1, 2, 3, 4, and 5, which is another sequence entirely.

By now, you should see the trick that is involved. Say you are given the sequence a, b, c and you want to make the next number to come out x, where x is any real number. First form the expression

$$(n - 1)(n - 2)(n - 3)$$

which will always have the value 0 for $n = 1$, $n = 2$, and $n = 3$. Then add to $(n - 1)(n - 2)(n - 3)$ some expression in n that produces a, b, and c when 1, 2, and 3 are substituted for n. (You should be able to show that this can always be done.) The fourth number produced by $(n - 1)(n - 2)(n - 3) +$ (the expression) is 6 greater than that expression produces because $(4 - 1)(4 - 2)(4 - 3) = 3 \cdot 2 \cdot 1 = 6$. Suppose the formula that gives a, b, and c gives d as the fourth number. Instead, you get $d + 6$.   However, $d + 6$ is not necessarily equal to x. To get x, you multiply the expression $(n - 1)(n - 2)(n - 3)$ by some number p such that $d + 6p = x$. Solving for p,

$$p = \frac{x - d}{6}$$

For example, suppose you want to generate the sequence

$$1,\ 2,\ 3,\ 5$$

Then it can be obtained by substituting 1, 2, 3, 4 in order in the formula

$$\frac{1}{6}(n - 1)(n - 2)(n - 3) + n$$

Incidentally, if you want to use this idea to confuse a test giver, it is much more mysterious if you carry out the multiplication. It is not obvious at all that

$$\frac{n^3}{6} - n^2 + \frac{17n}{6} - 1$$

is the correct expression to give

$$1,\ 2,\ 3,\ 5$$

when 1, 2, 3, 4, is substituted for n in order.

To eliminate such aberrations in sequences, a "formula" that gives any term of the sequence should be included in the statement of the sequence.

$$1,\ 2,\ 3,\ \ldots,\ n,\ \ldots$$

is the proper way to indicate the sequence of natural numbers in order. The "formula" for any term is simply n. And

$$1, 4, 9, \ldots, n^2, \ldots$$

is the proper way to indicate the sequence of squares of natural numbers. The "formula" is n^2. When you see such a "formula" in a sequence, it is called the *general term* of the sequence. By convention, the sequence is formed by replacing n in the general term by the sequence

$$1, 2, 3, \ldots, n, \ldots$$

one at a time in that order. (The circularity of this definition does not seem to bother anyone, although you will see later that perhaps it should.)

To wind up this part of the discussion with one more example, the sequence

$$1, 2, 3, \ldots, \frac{n^3}{6} - n^2 + \frac{17n}{6} - 1, \ldots$$

is a sequence that has

$$1, 2, 3, 5, 9$$

as its first 5 terms. (I leave it to the reader to find the general term for a sequence that has $1, 2, 3, 5, 4$ as its first 5 terms.

With the notion of a general term in mind, the exact meaning of the three dots is clearer. With a general term, you can find a term of the sequence for every natural number. Even though you do not know what the 10th or 137th term of a particular sequence is, you can calculate it. What is more, for every term of the sequence, there is another term that follows it. This is one kind of infinity. It is called a *countable infinity* (or a *denumerable* infinity). The reasons for this name will become clearer in Chapter 5.

INFINITY ONE STEP AT A TIME

There is a very interesting class of sequences that are formed by adding the terms of sequences, one more term each time, to get a new sequence. For example, if you add the terms of

$$1, \ 2, \ 3, \ \ldots, \ n, \ \ldots$$

taking one more each time, you get a sequence that starts out

$$1, \ (1 + 2), \ (1 + 2 + 3), \ (1 + 2 + 3 + 4)$$

or

$$1, \ 3, \ 6, \ 10$$

Shortly, you will learn the general term of this sequence. Look at another such pair of sequences first, however. If you add the terms of the sequence

$$1, \ 3, \ 5, \ 7, \ \ldots, \ 2n - 1, \ \ldots$$

in this way you get a sequence that starts out

$$1, \ 4, \ 9, \ 16$$

This *looks like* it might be the sequence

$$1, \ 4, \ 9, \ 16, \ \ldots, \ n^2, \ \ldots$$

But by now you should be suspicious. The next term might turn out to be 49. However, it is 25. Still, unless you can check all the terms of the sequence formed by adding the terms $1, 3, 5, 7, \ldots . 2n - 1,$ $\ldots$ (which you cannot do, since there are a countable infinity of them), you cannot know for sure that the next number after n^2 will be $(n + 1)^2$. What is to be done?

You need a way to show that the following equation is true for every natural number n.

$$1 + 3 + 5 + 7 \ldots + 2n - 1 = n^2$$

You cannot do it with ordinary algebra, because the "and so forth" and the general term at the end mean that this equation represents a countable infinity of equations.

What

$$1 + 3 + 5 + 7 + \cdots + 2n - 1 = n^2$$

means is that when $n = 1$,

$2n - 1 = n^2$ is true (and $2(1) - 1 = 1^2$ is true);

when $n = 2$,

$1 + 2n - 1 = n^2$ is true (and $1 + 2(2) - 1 = 2^2$ is true);

when $n = 3$,

$1 + 3 + 2n - 1 = n^2$ is true (and $1 + 3 + 2(3) - 1 = 3^2$ is true);

and so forth. Therefore, the first four of the countable infinity of equations "checks out." But you still have a countably infinite number that you have not checked.

The way out of this quandary is called *mathematical induction.* Mathematical induction is a method that can be used to prove a countable infinity of statements true in a finite number of steps. Here is how it works.

Step 1. First prove that the theorem is true for $n = 1$.

Step 2. Assume that the theorem is true for the first k natural numbers. You do not know what number k is, but you know k exists from Step 1.

Step 3. Use the assumption of Step 2 to prove that the theorem is true for the first $(k + 1)$ natural numbers.

If you have completed these steps correctly, mathematicians assure us, you have proved that the theorem is true for all natural numbers.

Now apply mathematical induction to

$$1 + 3 + 5 + 7 + \cdots + 2n - 1 = n^2$$

Step 1. $2n - 1 = n^2$ for $n = 1$ since $2 \cdot 1 - 1 = 1^2$ is true.

Step 2. Assume $1 + 3 + 5 + 7 + \cdots + 2k - 1 = k^2$ is true for some natural number k.

Step 3. Using your assumption, prove that the relationship is true for $k + 1$. Since k is assumed to be some specific natural number, the equation in Step 2 is an ordinary equation, not a countably infinite set. You can operate on it in the usual way. So it is possible to add $2k + 1$ to each side.

$$1 + 3 + 5 + 7 + \cdots + 2k - 1 + 2k + 1 = k^2 + 2k + 1$$

On the left side, you can rewrite the last two terms, $2k + 1$, as $2k + 2 - 1 = 2(k + 1) - 1$. On the right side, you can rewrite $k^2 + 2k + 1$ as $(k + 1)^2$.

$$1 + 3 + 5 + 7 + \cdots + 2k - 1 + 2(k + 1) - 1 = (k + 1)^2$$

Since this is the statement of the relationship for $k + 1$, you have completed the proof.

Still, it looks a bit like magic. The reasoning, however, is quite straightforward. The relationship is true for some number, since you have shown it to be true for $n = 1$. Then k is just some specific number for which the relationship is true. From the statement that it is true for k, you showed that it was true for $k + 1$. In other words, if it is true for 1, then it is true for 2; if it is true for 2, then it is true for 3; if it is true for 3, then it is true for 4; and so forth. Using finite reasoning, you have proved that

$$1 + 3 + 5 + 7 + \cdots + 2n - 1 = n^2$$

is true for all the countably infinite natural numbers that n can be.

Try another one. Look at the sequence of sums

$$1$$
$$1 + 2 = 3$$
$$1 + 2 + 3 = 6$$
$$1 + 2 + 3 + 4 = 10$$

You want to find a general term for this sequence $(1, 3, 6, 10, \ldots)$. Another way to say this, one that turns out to be very convenient, is to say that you want to find the sum of the *series*

$$1 + 2 + 3 + \cdots + n$$

A series is just a sequence whose terms are all summed (by replacing the commas with plus signs). The sum of a series is a number, but if the last number of the series is not specified, then the sum is the general term of the related sequence—in this case 1, 3, 6, 10, ... For example, the sum of $1 + 3 + 5 + \cdots + n$ is n^2. The related sequence is 1, 4, 9, $\cdots$, n^2, ...

With the squared numbers, it is easy to start by guessing the general term, because you are very familiar with squared numbers. These numbers, 1, 3, 6, 10, . . . , which are called *triangular numbers*, are not so familiar. Here is one way to guess the general term (or the sum of the series). You know that

$$1 + 3 + 5 + 7 + \cdots + 2n - 1 = n^2$$

Pretend that this is not a countably infinite set of equations and operate on it as if it were finite.

Notice that every odd number except 1 can be separated into two almost equal parts (that is, two parts whose difference is 1).

$$3 = 1 + 2$$
$$5 = 2 + 3$$
$$7 = 3 + 4$$
$$2n - 1 = (n - 1) + n$$

Replace the numbers in the equation above with the sums of these parts.

$$1 + (1 + 2) + (2 + 3) + (3 + 4) + \cdots + [(n - 1) + n] = n^2$$

Now separate this series into two parts as follows:

$$1 + 2 + 3 + 4 + \cdots + n + (1 + 2 + 3 + \cdots + n - 1) = n^2$$

It looks like you are close to the series you are looking for.

If you add n to the part in parentheses, the result is twice the triangular number series. Adding n to each side gives

$$1 + 2 + 3 + 4 + \cdots + n + (1 + 2 + 3 + \cdots + n - 1 + n) = n^2 + n$$

Since the left side is now twice the sum you are looking for, the sum must be

$$\frac{n^2 + n}{2}$$

Now you can use mathematical induction to find out whether or not this result—obtained by questionable means—is really true.

It is true for $n = 1$, since, when 1 is substituted for n, you get $1 = (1^2 + 1)/2$. Therefore, you can assume that for some number k, it is true.

$$1 + 2 + 3 + \cdots + k = \frac{k^2 + k}{2}$$

Add $k + 1$ to each side.

$$1 + 2 + 3 + \cdots + k + k + 1 = \frac{k^2 + k}{2} + k + 1$$

$$= \frac{k^2 + k}{2} + \frac{2k + 2}{2}$$

$$= \frac{k^2 + 2k + 1 + (k + 1)}{2}$$

$$= \frac{(k + 1)^2 + (k + 1)}{2}$$

Since this last result is what you set out to prove, the relationship is true for all natural numbers.

Now look at a more general case. Consider any sequence of numbers a, b, c, ..., X, ... where X represents the general term. You can use mathematical induction to show that all of the terms of this sequence are equal to a. Unlike the inductions just performed, however, this is clearly a fallacy.

Step 1. When the sequence has one term, you get $a = a$, which is always true.

Step 2. Assume that any set of k terms will have all of the terms equal to a for some natural number k.

Step 3. Now consider a set of $k + 1$ terms. From Step 2 you know that the first k terms are equal to a. Also, from Step 2, you know that the last k terms are equal to a. This can be shown as a diagram, where n is the $(k + 1)$th term.

$$\underbrace{a,\ a,\ a,\ \ldots,}_{k \text{ terms}}\ n$$

$$a,\ \underbrace{a,\ a,\ \ldots,\ n}_{k \text{ terms}}$$

From the bottom diagram, it is easy to see that, since the last k terms are equal to a and the first term is a, all of the terms are equal to a.

CAN YOU FIND THE FLAW?
Hint: Try this "proof" with some
very short sequences

Before discussing this fallacy, it is interesting to see it in another form. The theorem to be proved is "All billiard balls are the same color." You know that one billiard ball is the same color as itself. Assume some number of billiard balls are all the same color (which is safe, since you know that it is true for the number 1).

same color

Figure 2-1.

Now consider one additional billiard ball.

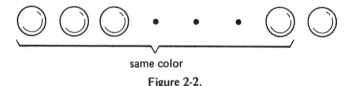

same color

Figure 2-2.

Since you know that the balls in braces are the same color for that number, you can move the braces over.

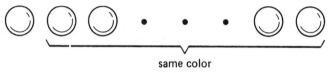

same color

Figure 2-3.

But the first ball in the row was already shown to be the same color as the others (except for the last in the row), so all balls in the row are the same color.   By the principle of mathematical induction, then, all billiard balls are the same color.

In this form, the problem should be clear. Consider the next simplest case to a single ball. With two balls, the diagrams look like this.

same color   same color

Figure 2-4.

Both diagrams represent true situations, but they do not imply that the two balls are the same color as each other. Induction fails because you cannot get from 1 to 2. (If you could get from 1 to 2, would the "proof" represent reality instead of fallacy?)

But, you might point out, the two earlier proofs by mathematical induction did not require looking at two terms of a sequence. That is true. The problem, however, with the two fallacies is that the reasoning used does not apply unless there are at least three terms of the sequence or three billiard balls. The reasoning used in the earlier proofs does not have this limitation.

Moving braces about may bother you since it is not a strictly mathematical operation. Can proof by mathematical induction fail when strictly mathematical operations are involved? Of course it can.

While you have applied mathematical induction principally to sequences, the method works for other mathematical entities as well. In the following example, induction is applied to a common function used in mathematics—the *maximum function*. The maximum function for two numbers is a mathematician's way of choosing the greater of two numbers. Thus, the maximum of 3 and 5 is 5; the maximum of 7 and 4 is 7. You can write this as follows:

$$MAX(3, 5) = 5$$
$$MAX(7, 4) = 7$$

To complete the definition of MAX, you need to define the maximum of a pair of equal numbers as $MAX(x, x) = x$, so $MAX(5, 5) = 5$, for example.

Now you are prepared to prove that for any number n if a and b are two natural numbers such that $MAX(a, b) = n$, then a is equal to b. (From the examples just given, it is easy to see that this proof is a fallacy.)

Step 1. Let $n = 1$. Then the "theorem" states that if $MAX(a, b) = 1$ for two natural numbers a and b, then $a = b$. Since the natural numbers do not include 0, the only pair of numbers that have 1 as their maximum is 1 and 1. Since $MAX(1, 1) = 1$ and $1 = 1$, the "theorem" is true for $n = 1$.

Step 2. Assume that k is some number for which the "theorem" is true. That is, if $MAX(a, b) = k$, then it is true that $a = b$. (From Step 1 it is clear that k exists.)

Step 3. Now consider the statement $MAX(p, q) = k + 1$, where p and $\dot{q}$ are two natural numbers. Define $x = p - 1$ and $y = q - 1$. Look at the function $MAX(x, y)$. Since both x and y are just 1 less than the numbers p and q, it is clear that $MAX(x, y)$ is equal to k. By Step 2, that means that $x = y$ is true. So $x + 1 = y + 1$ is true, or $p = q$. Thus, if $MAX(p, q) = k + 1$, then $p = q$, and the "theorem" is proved.

> **CAN YOU FIND THE FLAW?**
> **Hint**: Think of the numbers you
> are dealing with.

Since only mathematical operations are involved (no braces), this ought to be a legitimate proof. But the facts contradict it, so it is a fallacy.

It helps to look at specific numbers, just as you did in Chapter 1. The only specific numbers that you know to be in line with the theorem are 1, 1, and 1; that is, $MAX(1, 1) = 1$. So look at $MAX(p, q) = 1 + 1$, or 2. The natural numbers p and q such that $MAX(p, q) = 2$ are

$$MAX(1, 2) = 2$$
$$MAX(2, 1) = 2$$

and

$$MAX(2, 2) = 2$$

In Step 3 you assumed that 1 less than each of these numbers in the pairs would have the maximum 1. Here are those statements.

$$MAX(0, 1) = 1$$
$$MAX(1, 0) = 1$$

and

$$MAX(1, 1) = 1$$

All three statements are true, but the first two involve 0, which is not a natural number. In Step 1, the statement that $MAX(a, b) = 1$ implied that $a = b$ was true only because a and b were both natural numbers. Once you leave that assumption behind, the whole chain of reasoning is broken.

If you try to avoid this by changing Step 3 to eliminate subtraction, you get something like the following: Assume $MAX(a, b) = k$, then $MAX(a + 1, b + 1) = k + 1$. This is certainly true, but it does not prove the "theorem." For the "theorem" to be true, you would have to show that $\text{MAX}(p, q) = k + 1$, where p and q are any two natural numbers. For example, $\text{MAX}(a, b + 1) = k + 1$ is true if $\text{MAX}(a, b) = k$, but a is not equal to $b + 1$ when a is equal to b.

SOME SUM

In dealing with mathematical induction, you have been using a finite proof to cover a countable infinity of cases. Now you are going to shift gears slightly. Instead of dealing with a countable infinity of cases, each one of which is finite, you will deal with single cases that have a countable infinity of terms. This kind of infinity is usually handled very formally in mathematics courses, but since this is not a course in mathematics, you can take a relaxed approach to it. The need for some kinds of formality will emerge as you see how informal methods can lead you astray. (In fact, they led some of the greatest mathematicians of all time astray, so don't feel foolish.)

First, you are going to look at one more proof by mathematical induction of the same type considered previously. You are going to prove that

$$1 + x + x^2 + \cdots + x^n = \frac{1 - x^{n+1}}{1 - x}$$

Step 1. $1 = (1 - x)/(1 - x)$.

Step 2. Assume $1 + x + x^2 + \cdots + x^k = (1 - x^{k+1})/(1 - x)$ for some natural number k.

Step 3. Add x^{k+1} to each side of the equation in Step 2.

$$1 + x + x^2 + \cdots + x^k + x^{k+1} = \frac{1 - x^{k+1}}{1 - x} + x^{k+1}$$

$$= \frac{1 - x^{k+1}}{1 - x} + \frac{x^{k+1}(1 - x)}{1 - x}$$

$$= \frac{1 - x^{k+1} + x^{k+1} - x^{k+2}}{1 - x}$$

$$= \frac{1 - x^{(k+1)+1}}{1 - x}$$

This completes the proof.

Now consider what can happen if you continue this series without stopping at some particular n.

$$1 + x + x^2 + \cdots + x^n + \cdots$$

Is it possible to determine a sum for this series that has a countable infinity of terms? The answer turns out to be "yes and no." Look at some specific numbers for x.

$x = 2$ gives $1 + 2 + 4 + 8 + 16 + 32 + \cdots + 2^n + \cdots$

$x = 1$ gives $1 + 1 + 1 + 1 + 1 + 1 + \cdots + 1^n + \cdots$

$x = \dfrac{1}{2}$ gives $1 + \dfrac{1}{2} + \dfrac{1}{4} + \dfrac{1}{8} + \dfrac{1}{16} + \dfrac{1}{32} + \cdots + \left(\dfrac{1}{2}\right)^n + \cdots$

It would appear that the series for $x = 2$ becomes larger than any number you can name. In this case, the series does not have a sum. Also, the series for $x = 1$ becomes larger than any number you can name. For example, the sum is one million after a million terms, but the next million terms will bring it to a sum of two million, and so forth. The series for $x = 1/2$, however, has the possibility of having a sum. Look at the sequence formed by adding the terms of the series one more each time.

$$1 = 1$$

$$1 + \frac{1}{2} = 1\frac{1}{2}$$

$$1 + \frac{1}{2} + \frac{1}{4} = 1\frac{3}{4}$$

$$1 + \frac{1}{2} + \frac{1}{4} + \frac{1}{8} = 1\frac{7}{8}$$

$$1 + \frac{1}{2} + \frac{1}{4} + \frac{1}{8} + \frac{1}{16} = 1\frac{15}{16}$$

This sequence can be written as the following:

$$1, \; 1\frac{1}{2}, \; 1\frac{3}{4}, \; \ldots, \; 1 + \frac{n-1}{n}, \; \ldots$$

It is easy to see that no term of this sequence will ever be as great as 2, since $(n-1)/n$ can never be as great as 1. The further out in the sequence you go, however, the closer to 2 you get. The millionth term is 1-999,999/1,000,000, which is pretty close to 2, but the two millionth term is 1-1,999,999/2,000,000, which is even closer. When this kind of thing occurs, mathematicians call 2 the *limit* of the sequence. A sequence has a limit if there is some number such that for any term close to that number all the subsequent terms are closer still. The number that the terms are close to is the limit. If all of the terms are the same, as in the sequence 1, 1, 1, . . . , 1, . . . , then the limit is the same as one of the terms (for the sequence whose terms are all 1, the limit is 1).

Physical considerations suggest that the sum of an infinite series be defined as the limit of the sequence obtained by summing the terms of the series one more each time. Here is one example of a physical situation that suggests this definition. Say that a tortoise is traveling down the road at a speed of 1 kilometer per hour. At the end of the first hour, he travels 1 kilometer; at the end of the next half hour, he travels 1-1/2 kilometers; at the end of the next quarter hour, 1-3/4 kilometers, and so forth. The tortoise's progress is represented by the series

$$1 + \frac{1}{2} + \frac{1}{4} + \cdots + \left(\frac{1}{2}\right)^n + \cdots .$$

While this is an infinite series, it is clear to most folks that at the end of 2 hours the tortoise travels 2 kilometers. And 2 is the limit of the sequence of total distances as well as the sequence of total times. Therefore, one can write

$$1 + \frac{1}{2} + \frac{1}{4} + \cdots + \left(\frac{1}{2}\right)^n + \cdots = 2.$$

Here is another example of the limit of a sequence.

$$1, \frac{1}{2}, \frac{1}{4}, \ldots, \left(\frac{1}{2}\right)^n, \ldots$$

has the limit 0. The tenth term is $1/1024$, which is pretty close to 0, but every subsequent term is closer. All the terms after $(1/2)^n$ are closer to zero than $(1/2)^n$ is. In fact, it is easy to see that if x is greater than 1 (or less than -1) the sequence

$$1, \frac{1}{x}, \frac{1}{x^2}, \ldots, \frac{1}{x^n}, \ldots$$

has the limit 0. This is because the number x^{n+1} is always greater than x^n for any value of n. Another way to say the same thing is to say if x is less than 1 (and greater than -1) the sequence $1, x, x^2, \ldots,$ $x^n, \ldots$ always has the limit 0.

Now you are ready to use the notion of the limit of a sequence to find the sum of the infinite series

$$1 + x + x^2 + \cdots + x^n + \cdots$$

You have already proved that the general term of the associated sequence of sums is

$$\frac{1 - x^{n+1}}{1 - x}$$

For any value of x that is less than 1 (and greater than -1), the limit of the sequence $x, x^2, x^3, \ldots, x^{n+1}, \ldots$ must be 0 for the same reasons that the limit of $1, x, x^2, \ldots, x^n, \ldots$ is 0, even though these are different sequences. Now look at the sequence of terms found by summing the series one term at a time.

$$1, \ 1 + x, \ 1 + x + x^2, \ \ldots, \ \frac{1 - x^{n+1}}{1 - x}, \ \ldots$$

It helps to rewrite the general term as

$$\frac{1}{1 - x} - \frac{x^{n+1}}{1 - x}$$

For any particular value of x, the first fraction, $1/(1 - x)$, stays the same for any value of n. But if n gets bigger and if x is between -1 and 1, the numerator of the second fraction has a limit of 0 while the denominator does not change. Therefore, the second fraction also has a limit of 0. So,

$$1 + x + x^2 + \cdots + x^n + \cdots = \frac{1}{1 - x} \qquad (1)$$

when x is between -1 and 1. Similarly, when x is not between -1 and 1, the series does not have a sum at all.

Look at a couple of more series. From the one just given it is easy to derive

$$x + x^2 + x^3 + \cdots + x^{n+1} + \cdots = \frac{x}{1 - x} \qquad (2)$$

by subtracting 1 from each side of Equation (1) and rewriting 1 on the right side as $(1 - x)/(1 - x)$.

$$\frac{1}{1 - x} - \frac{1 - x}{1 - x} = \frac{1 - 1 + x}{1 - x} = \frac{x}{1 - x}$$

To find the sum of

$$1 + \frac{1}{x} + \frac{1}{x^2} + \cdots + \frac{1}{x^n} + \cdots$$

you can begin by substituting $1/x$ in Equation (1) for the sum of $1 + x + x^2 + \cdots + x^n + \cdots$, which gives

$$1 + \frac{1}{x} + \frac{1}{x^2} + \cdots + \frac{1}{x^n} + \cdots = \frac{1}{1 - \dfrac{1}{x}} \qquad (3)$$

This can be rewritten by multiplying numerator and denominator on the right side of Equation (3) by x to produce

$$1 + \frac{1}{x} + \frac{1}{x^2} + \cdots + \frac{1}{x^n} + \cdots = \frac{x}{x - 1} \qquad (4)$$

The eighteenth-century Swiss mathematician Leonhard Euler (1707–1783) was famous for his ability to calculate both numbers and sums of infinite series with astonishing facility. In fact, much of his best work was done during the 17 years that he was blind, a feat not unlike Beethoven's composing symphonies after he became deaf. Despite his great abilities, he often let algebraic logic overpower his sense of reality, especially when dealing with infinite series. The careful logic of limits was not developed in Euler's day. For example, he used the two series we have just been calculating as follows.

You know that

$$x + x^2 + x^3 + \cdots + x^n + \cdots = \frac{1}{1-x}$$

and

$$1 + \frac{1}{x} + \frac{1}{x^2} + \cdots + \frac{1}{x^n} + \cdots = \frac{x}{x-1}$$

Consider the sum of these two series. It is

$$\cdots + \frac{1}{x^2} + \frac{1}{x} + 1 + x + x^2 + \cdots = \frac{x}{1-x} + \frac{x}{x-1}$$

$$= \frac{x}{1-x} - \frac{x}{1-x}$$

$$= 0 \qquad (5)$$

This relation ought to be true for any x, but if x is positive, all of the terms of the infinite series will be greater than zero. How can the sum of an infinite number of positive numbers be zero?

> **CAN YOU FIND THE FLAW?**
> **Hint:** What is the right restriction to put on x in Equation (5)?

It may bother you to see the "and so forth" on both ends of the sequence. If so, just rearrange the terms to form

$$1 + \frac{1}{x} + x + \frac{1}{x^2} + x^2 + \frac{1}{x^3} + \cdots = 0$$

(Note that I have dropped the general terms from these series to save space, but they can be easily determined; no tricks!) The same problem with positive values of x occurs in this form. All of the terms of the series are positive, but the sum is zero.

Look at a specific number, however. Instead of the series that presents the fallacy, however, look at the first series you calculated

$$1 + x + x^2 + \cdots + x^n + \cdots = \frac{1}{1-x} \tag{1}$$

when x has the value 2. You get

$$1 + 2 + 4 + \cdots + 2^n + \cdots = \frac{1}{1-2} = -\frac{1}{2}$$

This result is clearly false. The sum of an infinite number of positive terms cannot be negative. But recall that when you derived your formula for the sum of the series, it was required that x be less than 1 (and greater than -1). The derivation of the formula breaks down for x equal to or greater than 1. It is not legitimate to use 2 as a replacement for x.

Notice that when x is any number other than 1, then if x is *not* between 1 and -1, then $1/x$ *is* between 1 and -1. Also, if x *is* between 1 and -1, then $1/x$ is *not*. Therefore, for any number other than 1, either the series

$$x + x^2 + x^3 + \cdots + x^n + \cdots$$

or the series

$$1 + \frac{1}{x} + \frac{1}{x^2} + \cdots + \frac{1}{x^n} + \cdots$$

will fail to have a sum. For the series that fails to have a sum, however, use of the formula produces a false result that is the negative of the correct result for the other series. Therefore, the sum of the two series, as calculated by their formulas, will be zero. Even though Euler was aware that working with series that do not have sums leads to false and paradoxical results, he apparently could not resist calculating with them.

Notice that if x is replaced by 1 (or -1) in either formula, the series do not have sums. But in each case, the formula for the sum gives your old friend 1/0. Euler believed that 1/0 was a number (infinity) which he represented as ∞. This was one belief that kept getting him into trouble with infinity.

It is not at all obvious what happens when x is negative in many series, for the terms in a series such as $1 + x + x^2 + \cdots + x^n + \cdots$ will be alternately positive and negative. Even though it has been noted that x cannot be less than -1 for this series to have a sum, perhaps this restriction is unnecessary. Look at a specific case. Let $x = -2$. Then the series becomes

$$1 - 2 + 4 - 8 + 16 - 32 + \cdots + (-2)^n + \cdots$$

Since some numbers are being added and some subtracted, it is at least thinkable that the series might have a sum.

Here is one way to find the sum. Call the sum S. Then

$$1 - 2 + 4 - 8 + \cdots = S$$

(omitting the general term for ease in reading). Notice that all of the terms are just the doubles of the terms immediately preceding, so you can rewrite the series as

$$1 - 2(1 - 2 + 4 - 8 + \cdots) = S$$

Then, the series in parentheses also has the sum S, so you know that

$$1 - 2S = S$$

or

$$1 = 3S$$

SO

$$S = \frac{1}{3}$$

But all of the terms of the series are integers. How can the sum be a number that is not an integer?

> **CAN YOU FIND THE FLAW?**
> **Hint**: Use the definition of the sum of a series.

Before unraveling this fallacy, it is instructive to look at two other methods of calculating the sum. You can put the parentheses in other places to get somewhat different results.

$$1 + (-2 + 4) + (-8 + 16) + \cdots$$

is the same series with the parentheses inserted a countable infinity of times. Completing the operations in the parentheses gives

$$1 + 2 + 8 + \cdots,$$

a series with all positive terms that has no sum.

On the other hand, you can insert the parentheses this way.

$$(1 - 2) + (4 - 8) + (16 - 32) + \cdots$$

which gives the following series when all of the operations in parentheses are completed:

$$-1 - 4 - 16 - \cdots$$

This is a series with all negative terms that has no sum.

The definition of the sum of a series is that it is the limit of the associated sequence. If you look at the sequence

$$1, \ 1 - 2, \ 1 - 2 + 4, \ \ldots$$

or

$$1, \ -1, \ 3, \ \ldots$$

you can get a better idea of what is happening. The fourth term is $1 - 2 + 4 - 8 = -5$; the fifth is $1 - 2 + 4 - 8 + 16 = 11$; the sixth is -21; the seventh, 43; and so forth. There is no number such that the terms of the sequence get close to it. Instead, the terms are alternately positive and negative (and farther apart with each succeeding pair of terms). When a infinite series does not have a sum, you can play all sorts of tricks with it.

Here is an easy example of the same kind to contemplate.

$$1 - 1 + 1 - 1 + 1 - 1 + \cdots$$

is equal to

$$1 - (1 - 1) - (1 - 1) - \cdots = 1$$

and is equal to

$$(1 - 1) + (1 - 1) + (1 - 1) + \cdots = 0$$

(But since the series does not have a sum, neither equation is true.)

The early explorers of infinite series, of course, struggled to develop definitions. You had the definition from the start. So it is not surprising that they tried hard to reconcile these conflicting results. Gottfried Wilhelm Leibniz (1646–1716), for example, suggested that

$$1 - 1 + 1 - 1 + \cdots$$

must have the value $1/2$, since that is the average of the two conflicting values obtained by placing the parentheses in different ways, as shown above. In fact, you can use the method of the last fallacy to "prove" that Leibniz was right. Let S equal the sum. Then

$$1 - (1 - 1 + 1 - 1 + \cdots) = S$$

or

$$1 - S = S$$

which has the solution $S = 1/2$. (But, of course, since $1 - 1 + 1 - \cdots$ has no sum, all of the reasoning above is false.)

Another mathematician of the time, trying to make some sense of this, wrote to Leibniz suggesting that since

$$(1 - 1) + (1 - 1) + (1 - 1) + \cdots = 0 + 0 + \cdots = 1/2,$$

this sequence explained how God created the world out of nothing. (God surely is 1.) He was serious!

If a series has a sum, it cannot lead to a fallacy, you might think. However, it can.

Consider the series

$$1 - \frac{1}{2} + \frac{1}{3} - \frac{1}{4} + \cdots + \frac{1}{n_{odd}} - \frac{1}{n_{even}} + \cdots$$

This series has a sum. You can show this as follows. First of all, since you have only my word that the series has a sum, you will work only with finite series. Consider using parentheses to write the finite series of the first n terms as follows.

$$\left(1 - \frac{1}{2}\right) + \left(\frac{1}{3} - \frac{1}{4}\right) + \left(\frac{1}{5} - \frac{1}{6}\right) + \cdots + \left(\frac{1}{n_{odd}} - \frac{1}{n_{even}}\right)$$

Notice that each of the expressions in parentheses is positive. In general, each is of the form $1/k - [1/(k + 1)]$. Since $k + 1$ is always larger than k, $1/(k + 1)$ is always smaller than $1/k$, so $1/k - [1/(k + 1)]$ is always positive. Call this *Result Number 1*. Another way of stating Result Number 1 is that for any n the sum of the finite series is a positive number; and for $n + 1$ the sum is a larger positive number.

Result Number 1 might sound like an implication that the infinite series has no sum, but in fact the same result could be stated for the series

$$1 + \frac{1}{2} + \frac{1}{4} + \cdots + \frac{1}{2^n}$$

which you know has a sum at infinity.

Now move the parentheses to form the finite series

$$1 - \left(\frac{1}{2} - \frac{1}{3}\right) - \left(\frac{1}{4} - \frac{1}{5}\right) - \cdots - \left(\frac{1}{n_{\text{even}}} - \frac{1}{n_{\text{odd}}}\right)$$

By the same reasoning (in reverse) as above, all the terms in parentheses have to be negative. Therefore, the sum of this finite series, for any value of n, must be less than 1, since it can be written as 1 minus a number of terms. This is *Result Number 2*.

Combining Result Number 1 and Result Number 2 shows that for any n the finite series is positive but less than 1. What is more, from Result Number 1, if n gets bigger, the sum gets bigger. Since in

$$\left(1 - \frac{1}{2}\right) + \left(\frac{1}{3} - \frac{1}{4}\right) + \cdots + \left(\frac{1}{n_{\text{odd}}} - \frac{1}{n_{\text{even}}}\right)$$

the number in the first parentheses $(1 - 1/2)$ is $1/2$, the sum of the finite series will always be between $1/2$ and 1. If you think about this a while, you can see that there must be some number between $1/2$ and 1 that is the limit for this finite series

$$1 - \frac{1}{2} + \frac{1}{3} - \frac{1}{4} + \cdots + \frac{1}{n_{\text{odd}}} - \frac{1}{n_{\text{even}}} + \cdots$$

While this result should be intuitively clear, actually showing that this is the case (or showing what the limit is) is beyond the scope of this book.

(In fact, the limit is a number near 0.6931. You might think, from the above paragraph, that the limit would be 1. Looking more closely at the finite series in the version

$$1 - \left(\frac{1}{2} - \frac{1}{3}\right) - \cdots - \left(\frac{1}{n_{\text{even}}} - \frac{1}{n_{\text{odd}}}\right)$$

however, will show you that as n gets bigger, the sum of the finite series is farther from 1, not closer. It would have to get closer and closer to 1 for 1 to be the limit.)

Now that you are sure that the infinite series has a sum, you can call that sum S and do some calculations with it. For example, multiply the series by 1/2 and you get

$$\frac{1}{2} - \frac{1}{4} + \frac{1}{6} - \frac{1}{8} + \cdots = \frac{1}{2}S$$

(This will be easier to follow if the general terms are omitted, although the results are the same if they are included.) Now add $(1/2)S$ to S. This is easiest to do if you set up the addition like an arithmetic problem and align fractions with the same denominator.

$$1 - \frac{1}{2} + \frac{1}{3} - \frac{1}{4} + \frac{1}{5} - \frac{1}{6} + \frac{1}{7} - \frac{1}{8} + \cdots = S$$
$$+ \quad \frac{1}{2} \quad - \frac{1}{4} \quad + \frac{1}{6} \quad - \frac{1}{8} + \cdots = +\frac{1}{2}S$$
$$\overline{\quad 1 + 0 + \frac{1}{3} - \frac{1}{2} + \frac{1}{5} + 0 + \frac{1}{7} - \frac{1}{4} + \cdots = \frac{3}{2}S}$$

The infinite series for $(3/2)S$ may look vaguely familiar. With the zeros omitted, it is

$$1 + \frac{1}{3} - \frac{1}{2} + \frac{1}{5} + \frac{1}{7} - \frac{1}{4} + \cdots$$

In fact, if you carry it out a few more terms, it is recognizable as a rearrangement of

$$1 - \frac{1}{2} + \frac{1}{3} - \frac{1}{4} + \frac{1}{5} - \frac{1}{6} + \cdots$$

whose sum is S. Therefore, the series whose sum is $(3/2)S$ is the same as the series whose sum is S. But $S = (3/2)S$ is true if and only if $S = 0$, whereas you have shown that S is greater than $1/2$.

CAN YOU FIND THE FLAW?
Hint: Look at the series of all positive terms and all negative terms.

It should be noted that this particular result gave the eighteenth-century mathematicians heartburn. Unlike the fallacies discussed earlier in this book, the mathematicians who discovered it had no idea what was wrong—but they knew that it was something.

In fact, S can be shown to be any amount you like. For example, multiply the terms of the series by 2 this time.

$$2 - 2 \cdot \frac{1}{2} + 2 \cdot \frac{1}{3} - 2 \cdot \frac{1}{4} + 2 \cdot \frac{1}{5} - 2 \cdot \frac{1}{6} + \cdots = 2S$$

or

$$2 - 1 \quad + \frac{2}{3} \quad - \frac{1}{2} \quad + \frac{2}{5} \quad - \frac{1}{3} \quad + \cdots = 2S$$

This infinite series contains a number of terms with the same denominator. If you group these with parentheses, you get

$$(2 - 1) - \frac{1}{2} + \left(\frac{2}{3} - \frac{1}{3}\right) - \frac{1}{4} + \left(\frac{2}{5} - \frac{1}{5}\right) - \frac{1}{6} + \cdots = 2S$$

or

$$1 - \frac{1}{2} + \frac{1}{3} - \frac{1}{4} + \frac{1}{5} - \frac{1}{6} + \cdots = 2S$$

But, once again, the series is the one that has the sum S, so $S = 2S$, as well as $(3/2)S$.

Although this series (and others like it) were extensively discussed by Euler, it was not until 1854 (more than 50 years after Euler published his thoughts on such series) that Bernhard Riemann (1826–1866) proved that these kinds of series could produce any sum whatsoever when rearranged. Although they have a well-defined sum when presented in their original form, any manipulation that involves rearranging an infinite number of terms can make the series as meaningless as an infinite series without a sum.

The essence of Riemann's method is to look at the related series that has all positive terms. Consider

$$1 + \frac{1}{2} + \frac{1}{3} + \frac{1}{4} + \cdots + \frac{1}{n} + \cdots$$

Does this series have a sum?

It is easy to see that the sequence

$$1, \frac{1}{2}, \frac{1}{3}, \frac{1}{4}, \ldots, \frac{1}{n}, \ldots$$

has the limit 0. But the question of whether or not the *series* has a sum is answered by whether or not a different sequence has a limit. That sequence starts out

$$1$$

$$1 + \frac{1}{2} = 1\frac{1}{2}$$

$$1 + \frac{1}{2} + \frac{1}{3} = 1\frac{5}{6}$$

$$1 + \frac{1}{2} + \frac{1}{3} + \frac{1}{4} = 2\frac{1}{12}$$

$$1 + \frac{1}{2} + \frac{1}{3} + \frac{1}{4} + \frac{1}{5} = 2\frac{17}{60}$$

It is not easy to tell whether or not the sequence

$$1, 1\frac{1}{2}, 1\frac{5}{6}, 2\frac{1}{12}, 2\frac{17}{60}, \ldots$$

has a limit, nor is it easy to find a general term for the sequence. One way to get around these problems is to look at an easier sequence—one that is better behaved. The easiest fractions to work with are the powers of 1/2: 1/4, 1/8, 1/16, and so forth. Replace each term in the original series with the greatest power of 1/2 that is less than the term itself. (Remember this language; similar language returns to haunt you in a later chapter.) This procedure produces the series

$$\frac{1}{2} + \frac{1}{4} + \frac{1}{4} + \frac{1}{8} + \frac{1}{8} + \frac{1}{8} + \frac{1}{8} + \frac{1}{16} + \cdots$$

because 1/2 is less than 1, 1/4 is less than 1/2 and 1/3, 1/8 is less than 1/4, 1/5, 1/6, 1/7 and so forth. Since each term of this series is less than the corresponding term of the original series, if the new series does not have a limit, then the original does not. That is, if the sum of the new series increases "to infinity," so does the original one.

Now look at the first terms of the sequence of sums taken one term at a time.

$$\frac{1}{2}$$

$$\frac{1}{2} + \frac{1}{4}$$

$$\frac{1}{2} + \left(\frac{1}{4} + \frac{1}{4}\right) = \frac{1}{2} + \frac{1}{2}$$

$$\cdot$$
$$\cdot$$
$$\cdot$$

$$\frac{1}{2} + \left(\frac{1}{4} + \frac{1}{4}\right) + \left(\frac{1}{8} + \frac{1}{8} + \frac{1}{8} + \frac{1}{8}\right) = \frac{1}{2} + \frac{1}{2} + \frac{1}{2}$$

It is easy to see that the first term is 1/2, after 2 more terms the sum is 1/2 + 1/2, after 2^2 more terms the sum is 1/2 + 1/2 + 1/2, after 2^3 more terms the sum is 1/2 + 1/2 + 1/2 + 1/2, and so on. Although the sequence increases slowly, it will eventually get bigger than any number you can name. So it has no limit.

By the earlier argument, this implies that

$$1 + \frac{1}{2} + \frac{1}{3} + \frac{1}{4} + \cdots + \frac{1}{n} + \cdots$$

has no sum.

This was a long sidetrack away from the original problem, which was to explain the peculiar behavior of

$$1 - \frac{1}{2} + \frac{1}{3} - \frac{1}{4} + \cdots + \frac{1}{n_{\text{odd}}} - \frac{1}{n_{\text{even}}} + \cdots$$

which does have a sum. But look what happens when you start rearranging the terms. Put all positive terms first.

$$1 + \frac{1}{3} + \frac{1}{5} + \cdots + \frac{1}{n_{odd}} + \cdots - \frac{1}{2} - \frac{1}{4} - \frac{1}{6} - \cdots - \frac{1}{n_{even}} - \cdots$$

For this series to have a sum in this form, it is necessary that the sum of the positive terms exists and that the sum of the negative terms exists. Then, the sum of the series is the sum of the positive and negative numbers. But you can use the same trick with powers of 1/2 as was used for

$$1 + \frac{1}{2} + \frac{1}{3} + \frac{1}{4} + \cdots + \frac{1}{n} + \cdots$$

to show that neither the positive "half" nor the negative "half" has a sum. The positive "half" must be greater than

$$\frac{1}{2} + \frac{1}{4} + \left(\frac{1}{8} + \frac{1}{8}\right) + \left(\frac{1}{16} + \frac{1}{16} + \frac{1}{16} + \frac{1}{16}\right) + \cdots$$

or

$$\frac{1}{2} + \frac{1}{4} + \frac{1}{4} + \frac{1}{4} + \cdots + \frac{1}{4} + \cdots$$

This series has no sum, so the positive "half" has no sum.

Similarly, the negative "half" must be less than

$$-\frac{1}{2} - \frac{1}{4} - \frac{1}{4} - \frac{1}{4} - \cdots - \frac{1}{4} - \cdots$$

Since this series has no sum, the negative "half" has no sum either.

Although the rearranged series has no sum, the original series does. This seemingly paradoxical situation can occur because in the original arrangement, the amounts subtracted each time bring the related sequence of partial sums closer and closer to a particular number (around 0.6931). But if you add a finite number of terms in a different order and subtract a finite number of terms in a different order, although you still get the same number of terms, the sequence of partial sums can change all over the place. To give a specific example, the related sequence for

$$1 - \frac{1}{2} + \frac{1}{3} - \frac{1}{4} + \frac{1}{5} - \frac{1}{6} + \cdots$$

starts off

$$1, \ \frac{1}{2}, \ \frac{5}{6}, \ \frac{7}{12}, \ \frac{47}{60}, \ \frac{37}{60}, \ \cdots$$

and has a limit, but

$$1 + \frac{1}{3} - \frac{1}{2} + \frac{1}{5} + \frac{1}{7} - \frac{1}{4} + \cdots$$

has the related sequence

$$1, \ \frac{4}{3}, \ \frac{5}{6}, \ \frac{31}{30}, \ \frac{247}{210}, \ \frac{389}{420}, \ \cdots$$

which does not have a limit.

Riemann's test for whether or not a series that has both positive and negative terms converges is to change the signs of all the negative terms to positive. If the new series with all positive terms has a sum, then rearranging the terms in the original series will not make a difference. If, however, the new series of all positive terms does not have a sum, then rearrangement may produce a series with no sum. Thus, you can fiddle with

$$1 - \frac{1}{2} + \frac{1}{4} - \frac{1}{8} + \cdots + \frac{1}{(-2)^n} + \cdots$$

because

$$1 + \frac{1}{2} + \frac{1}{4} + \frac{1}{8} + \cdots + \frac{1}{2^n} + \cdots$$

has a sum. But you are not allowed to mess with

$$1 - \frac{1}{2} + \frac{1}{3} - \frac{1}{4} + \cdots + \frac{1}{n_{\text{odd}}} - \frac{1}{n_{\text{even}}} + \cdots$$

because

$$1 + \frac{1}{2} + \frac{1}{3} + \frac{1}{4} + \cdots + \frac{1}{n} + \cdots$$

does not have a sum. No matter how you rearrange the parts of a series that meets the Riemann test, even if you put all the positive terms first, the series continues to have a sum (and it stays the same).

A PARADOX FOR PITCHMEN

Although this chapter is concerned primarily with reasoning gone wrong, it is the best place to discuss an example of correct reasoning that results in a paradox. The paradox is of the type that mathematicians believe to be true—that is, the reasoning has been carefully examined, the paradox does not contradict any known result of mathematics or the real world, but the conclusion feels wrong. It goes against intuition. Nonetheless, the conclusion is accepted as true. The reason for discussing it here is that it is a direct result of the kind of thinking you have been doing about infinite series.

In the eighteenth century, one of the centers of mathematics was the Russian capital of St. Petersburg. This was not because there were a lot of great Russian mathematicians at that time; rather, an enlightened policy towards mathematics by the rulers of Russia brought great mathematicians from all over Europe to St. Petersburg. Among these were Nicholas and Daniel Bernoulli, two members of the famous Swiss family of mathematicians. While they were both there, they developed, in conversations, what has come to be know as the *Petersburg paradox*. It is a paradox that relates probability notions to infinite series. Another mathematician of the time, Jean Le Rond d'Alembert (whose name was based on the name of the church in Paris upon whose steps he had been abandoned by his aristocratic mother), called the Petersburg paradox a scandal. In fact, d'Alembert felt that something had to be wrong with probability theory for such a paradox to have occurred.

Here is a version of the Petersburg paradox. Simon J. Montague is running a booth at a carnival. There is an automatic coin-tossing machine that is absolutely fair. For a fee, you can play the game with Simon. The machine tosses the coins quickly and reports whether the result is heads or tails. You play until heads appears. If it is on the first toss, Simon will give you a dollar; if it is on the second toss, he will give you 2 dollars; on the third toss, 4 dollars; on the fourth toss, 8 dollars; and if heads does not appear until the nth toss, you will receive 2^{n-1} dollars.

You are attracted to this game, provided the fee to play is not too steep. Simon asks you if you know any mathematics. When you tell him that you know a little, he offers the following analysis of the game (before telling you what the fee is):

"The fee ought to be based upon your expectation of winning your money back, my boy. If there were just one coin toss, you would have a probability of 1/2 of winning, so an appropriate fee would be 50 cents. In other words, we would both stand a 50:50 chance of making four bits. If the toss is heads, I pay you a buck and you make fifty cents. If the toss is tails, I keep your fifty cents. Clearly, if we played that way over and over, it would be even in the eyes of Lady Luck."

Before you can protest that Lady Luck is supposed to be blind, Simon proceeds: "But that's not how we play the game. The machine continues until heads shows. The chance that heads shows on the second toss is 1/4, since the probabilities of independent tosses are multiplied. You did say you know some mathematics, didn't you? So if we agreed to stop the game after two tosses, an appropriate fee would be a dollar. Your chances of making a dollar and my chances of making one would be equal. Here are the combinations.

First toss:	H	I pay you $1.00. We're even.
	T	We toss again.
Second toss:	H	I pay you $2.00, you make $1.00.
	T	Game ends, I make $1.00.

"The way to calculate this, in case you don't know it, sonny, is as follows. The appropriate fee is the sum of the probabilities of winning multiplied by the payoffs. In other words, for a game ending in two tosses, it is

$$\frac{1}{2}(1) + \frac{1}{4}(2) = \frac{1}{2} + \frac{1}{2} = 1$$

For three tosses, it becomes

$$\frac{1}{2}(1) + \frac{1}{4}(2) + \frac{1}{8}(4) = \frac{1}{2} + \frac{1}{2} + \frac{1}{2} = 1\frac{1}{2}$$

"BUT THE MACHINE DOES NOT STOP UNTIL IT TOSSES HEADS! The appropriate fee then is

$$\frac{1}{2}(1) + \frac{1}{4}(2) + \frac{1}{8}(4) + \cdots + \frac{1}{2^n}(2^{n-1}) + \cdots$$

or

$$\frac{1}{2} + \frac{1}{2} + \frac{1}{2} + \cdots + \frac{1}{2} + \cdots$$

This series becomes larger than any number you can name. Therefore, the appropriate fee would be an infinity of dollars.

"But I know you don't have that much, so tell ya what I'm gonna do. I will let you play this game with me for just ten bucks. Yes sir, ten dollars will get you into the game."

Impressed by the mathematics, you agree to play. However, after a dozen games you find that you are out about $60 and broke.

> **CAN YOU FIND THE FLAW?**
> **Hint:** How many games would you
> have to play to be sure of at
> least breaking even?

I based the report that you lost about $60 in a dozen games on an experiment that Comte de Buffon (1707–1788) performed. He played the game 2,084 times with an actual coin. He found that, in this version of the game, he would have been paid $10,057. There-fore, a fee for the game that would have been "fair," in the sense that he would have broken even or nearly so after a great many games, would have been about $5, since 5 × 2,084 is about 10,057. If he had paid a fee of $10, he would have lost about $10,000.

No one has a really good explanation for this paradox. Simon J. Montague's reasoning is sound. However, you had better avoid his game.

Here is one way of looking at the paradox (there are others). Montague does not have an unending source of money, and neither do you. Say that at any point in the game, the most money that Montague can lay his hands on is 2^{n-1} dollars. Then you should pay $n/2$ dollars to have an even chance of winning. Say, n is 10. Then Montague has $512 (because $2^9 = 512$) and you have to pay $5 a game to play. But the game stops if you "break the bank." Notice that Montague's limit—he won't pay on more than $512—makes this a fair game. With no limit, however, all bets are off.

Another way of looking at your chances is to note that you will lose money on a $10 bet if heads comes up before the fifth toss. The probability of this happening is 15/16, so you only have 1 chance in 16 of making any money, which is not very good odds.

Look at a related gambling situation. Sometimes gamblers start doubling their bets. The idea behind this strategy is that there is no way the gambler can lose. Say that the game is one in which you are paid $5 for every dollar you bet if you win, but lose your bet if you lose. (If you win on a bet of $2, the house takes your $2 and gives you $10.) Suppose you have a long losing streak, but bet as follows:

You bet	Total loss	Possible gain on this bet
1	1	4
2	3	8
4	7	16
8	15	32
16	31	64
.	.	.
.	.	.
.	.	.
2^n	$2^{n+1} - 1$	2^{n+2}

Since 2^{n+2} is always bigger than $2^{n+1} - 1$, it is clear that at some point, when you finally win, you will be ahead. On the other hand, if you run out of money before you win, you will have lost $2^{n+1} - 1$ dollars. Unless you have a lot of cash to start with, the doubling strategy does not pay off very well. In fact, for it to work, you need always to be able to come up with the cash. For example, if you lose 10 times in a row, you need \$1,024 for the bet, you have already lost \$2,047, and you stand to win (on that bet) \$4,096 if you win, so if you win you are ahead \$1,025 and should quit now!

Both the Petersburg paradox and the doubling gambling strategy rely on infinite sums of money. While these are countable infinities, you have seen how easy it is to reason incorrectly about series that have no sums. Perhaps the fundamental problem with the Petersburg paradox is that it treats a series that has no sum as if it had an infinite sum. If you look at any finite stopping place, there is no paradox. It is only when you assume that sum to be infinite that a problem arises.

3
Using a Wrong Idea to Find Truth

... there is no evidence supporting the belief in the existential character of the totality of all natural numbers .... The sequence of numbers which grows beyond any stage already reached by passing to the next number is a manifold of possibilities open towards infinity; it remains forever in the status of creation, but is not a closed realm of things existing in themselves. That we blindly converted one into the other is the true source of our difficulties ...

Hermann Weyl

So far you have observed problems that arise in understanding the real world by experiment (Aristotle's circle paradox), in dealing with algebra (De Morgan's paradox), in understanding real events through logic (the Swedish civil-defense exercise paradox), in understanding infinity (Euler's paradox), and in understanding how chance works (the Petersburg paradox). But you have not seen any flaws in geometry. Is this because, in the words of Edna St. Vincent Millay, "Euclid alone has looked on beauty bare?"

Geometry, after all, is the first branch of mathematics in which all of the results are supposedly derived by simple rules of logic from a small set of self-evident postulates. You learned all about it in high school. Euclid's geometry has withstood the test of time. Even though there are non-Euclidean geometries around, nothing is wrong with Euclid's geometry.

Maybe.

DRAWING AN INSIDE STRAIGHT

A lot of the easy part of Euclid's geometry has to do with triangles. Look at any triangle.

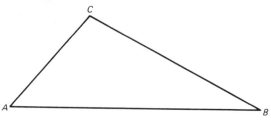

Since this is supposed to be *any* triangle, you cannot assume that any of the sides are equal. In particular, you cannot assume that $AC = BC$. If $AC = BC$ were true, then the triangle would be *isosceles*—and an isosceles triangle is not just any old triangle.

It is hard to prove something about a triangle just in isolation. So, draw a few lines on the triangle to give yourself something to work with. For example, bisect angle C. Draw the perpendicular bisector of side AB. These two lines will have to meet at a point, which you can label D.

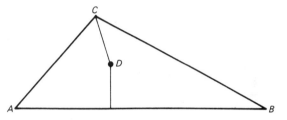

From point D you can draw lines to A and B and you can drop perpendiculars to sides AC and BC. Label the points on AC and BC where the perpendiculars meet at E and F.

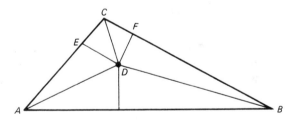

Now you have an interesting figure with all sorts of properties. For example, triangles CDE and CDF are both right triangles, since DE is perpendicular to AC and DF is perpendicular to BC. Also, $\angle ECD$ is equal to $\angle FCD$, since CD is the bisector of $\angle ACB$. Since $\triangle CDE$ and $\triangle CDF$ also have a side in common, you can show that they are the same in size and shape. In other words, $\triangle CDE$ and $\triangle CDF$ are *congruent*, as represented by

$$\triangle CDE \cong \triangle CDF$$

where $\cong$ means "is congruent to."

There are two ways to prove these two triangles congruent. The long way is to show that $\angle CDE = \angle CDF$, which shows that $\triangle CDE$ has two angles and the included side equal to two angles and the included side of $\triangle CDF$. This is proof by angle-side-angle (as it is usually called in high-school geometry).

The shorter way is to use one of the special congruence theorems for right triangles. If the hypotenuse and one acute angle of one right triangle are equal to the hypotenuse and one acute angle of another right triangle, the triangles are congruent. Since $CD = CD$ and $\angle ECD = \angle FCD$, the triangles are congruent. In congruent triangles, the corresponding parts are equal, so $DE = DF$.

You can also show that $DA = DB$. This relationship holds because D is on the perpendicular bisector of AB. Now for right triangles, which is what $\triangle DEA$ and DFB are, the equalities $DA = DB$ and $DE = DF$ are all you need to prove congruence. Two right triangles are congruent if the hypotenuse and one other side of one of them are equal to the hypotenuse and the corresponding side of the other. Therefore,

$$\triangle DEA \cong \triangle DFB$$

Since corresponding parts are equal,

$$EA = FB$$

But from

$$\triangle CDE \cong \triangle CDF$$

you also know that

$$CE = CF$$

By adding, you can obtain

$$EA + CE = FB + CF$$

which implies that

$$AC = BC$$

In that case, the triangle is isosceles. Since you explicitly started with the assumption that $\triangle ABC$ was a general triangle, you have proved that all triangles are isosceles.

If you carefully draw a triangle that is not isosceles, draw the bisector of one angle and the perpendicular bisector of the opposite side, you get a drawing that will look like the following:

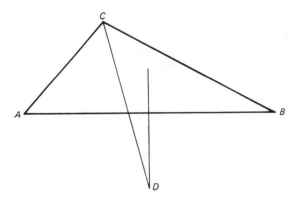

Following the rest of the instructions produces a diagram like this:

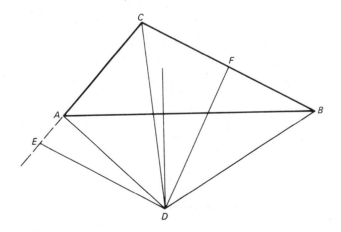

Clearly, this situation is totally different from the one on which the proof is based. And, if you make a series of careful drawings,

you will always find that, unless the triangle is isosceles, with $AC = CB$, point D will *always* lie outside of $\triangle ABC$. (What happens when $AC = CB$? Can you prove it?)

Since Euclid did not offer any postulates about outside and inside, it is difficult to *prove* that D will always be outside the triangle. There is one fairly easy proof, however. You have just been through it.

Here it is again, in case you missed it the first time.

Step 1. Assume $\triangle ABC$ is not an isosceles triangle ($AC \neq BC$, where $\neq$ means "is not equal to").

Step 2. Assume that CB, the bisector of $\angle C$, meets the perpendicular bisector of AB at a point *inside* $\triangle ABC$.

Step 3. Prove $\triangle ABC$ is isosceles as before. (There is nothing wrong with this proof if D is inside $\triangle ABC$.)

Step 4. Therefore, unless $\triangle ABC$ is isosceles, point D is not inside $\triangle ABC$.

This proves that point D cannot be inside the triangle unless $\triangle ABC$ is isosceles. You still need to prove that point D exists. For example, the bisector of $\angle C$ and the perpendicular bisector of AB could be parallel—in which case they would never meet.

Here, however, is a proof to show that they meet when $\triangle ABC$ is not isosceles.

Step 1. $\triangle ABC$ is not isosceles ($AC \neq BC$).

Step 2. Assume that the bisector of $\angle C$ is parallel to the perpendicular bisector of AB.

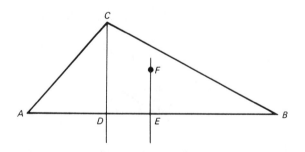

Step 3. Since FE is perpendicular to AB and CD is parallel to EF, then CD is perpendicular to AB. $\angle CDA = \angle CDB$, because

both are right angles. $\angle ACD = \angle BCD$, because CD bisects $\angle C$. This means $\angle A$ must equal $\angle B$. But if $\angle A = \angle B$, the triangle is isosceles.

Step 4. Since we know $\triangle ABC$ is not isosceles, then the bisector of $\angle C$ must meet the perpendicular bisector of AB in a point.

This completes the proof that if a triangle is not isosceles, the bisector of one angle meets the perpendicular bisector of the opposite side in a point outside the triangle.

You may have noticed that this is a peculiar form of proof. You set up the situation so that the opposite of what you want to prove is assumed. Then you prove that this contradicts some other basic feature of the problem. Therefore, since both cannot be true, the assumption you made must be false. Thus its opposite must be true. This form of proof is called *indirect proof.*

The indirect proof relies on a principle of logic called the *law of the excluded middle.* Basically, the law of the excluded middle says that, for any statement A, either A is true or not A is true, but not both. Later in this chapter, you will encounter an argument that the law of the excluded middle is "not operative" all of the time, but for now you can assume that it always works.

Look at another example. In this case, start with two intersecting circles of different sizes. These meet at points A and B.

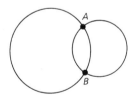

Suppose you draw a diameter of each circle from one of the points of intersection, say A. These diameters will meet the circles at points C and D. Connect points C and D with a line.

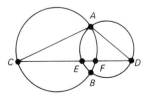

Call the points where CD intersects the circles E and F. Now draw lines from A to both E and F.

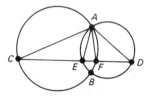

Since AC is a diameter, $\angle CFA$ is a right angle. Similarly, since AD is a diameter, $\angle AED$ is a right angle. Therefore, you have shown that from a point not on a line (A) there can be two perpendiculars to the line (CD). However, you remember that only one perpendicular can be drawn to a line from a point.

CAN YOU FIND THE FLAW?

Hint: Translate the fallacy into an indirect proof.

The only way that this result could be true is if B, E, and F were all the same point. So, the paradox becomes the basis for the following statement: if two circles intersect at two points and diameters are drawn from one point of intersection across the circles, the line connecting the other ends of the diameters must pass through the other point of intersection.

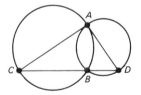

The paradox just given shows that the line cannot pass through the circles in the region between A and B. Could CD pass below B?

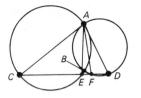

In this case, the same argument applies. *AE* and *AF* would be different perpendiculars to the same line. Since this cannot be, *CD* must always pass through point *B*.

The material presented so far in this chapter ought also to remind you not to depend on diagrams in your reasoning. While diagrams are usually helpful, they can also be quite misleading.

BEING AND NONBEING

The three indirect proofs given so far are slightly different from each other. For the two circles, you proved that a particular point lies on a particular line. This is quite like most proofs in geometry. For the general triangle, you proved that a particular point lies outside the triangle, but you could not pinpoint where it was beyond that: just "outside." Along the way, however, you had to prove that the point existed. Mathematicians call this last kind of proof an *existence proof*. An existence proof establishes the existence of some entity without telling you how to find it. Existence proofs are one of the backbones of mathematics.

Here is an example of an easy existence proof (that is not an indirect proof). The statement to be proved is that there are two people with the same number of hairs on their heads.

Proof: No one has as many as a billion hairs on their heads. There are more than a billion people on earth (quite a few more). Start with the person with the fewest hairs on his or her head. Check to see if there is another person with the same number. If so, your proof is complete. If not, move on to the person with the next fewest hairs on his or her head and repeat the process. You have more people than hairs, so you will have to find a match before you get to the billionth person. End of proof.

Existence proofs are a little unfilling (like Chinese meals are reputed to be) and an hour later many mathematicians want another proof— a *constructive proof* that will show exactly which people have the same number of hairs or will at least locate two such people. Nevertheless, a lot of mathematics consists of existence proofs with no corresponding constructive proof.

Perhaps less unsettling is the *nonexistence proof.* Clearly, this is a proof that some entity does not exist. For example, there is no uninteresting natural number, which can be shown by an indirect proof. In other words, you can prove the nonexistence of an uninteresting natural number.

Some numbers are interesting. For example, 1 is interesting because it is the first natural number, 2 is interesting because it is the only even prime, 3 is interesting because it is the first odd prime, and so forth. But what about 137 or 4,329? Consider the set of all uninteresting numbers. This set must have a least member. But that number must be interesting since it is the first uninteresting number. Since this is a contradiction, there is no uninteresting natural number.

It was a nonexistence proof that led to the first great crisis in mathematics. This proof was advanced by the followers of Pythagoras around 500 B.C. It was a true paradox for them (and they tried to keep it a secret for a while, since it undermined their basic beliefs), but today most mathematicians think of it as just another theorem (if they think of it at all). It is usually taught in high school, and therefore falsely viewed as unimportant.

A bit of background is needed to understand why the Pythagoreans were so disturbed. Their leader, Pythagoras, taught that all things were based on number—by which he meant the natural numbers. Measurements that were not natural numbers could be found as the ratios of two natural numbers—what are called fractions in elementary school and *rational numbers* by mathematicians. For example, if you measure a line and it comes to a little more than one unit, you can identify its length as the ratio of two numbers—say 5 to 4 (or 1-1/4 units). If the length is a little less than 5 to 4, you could measure the length as 19 to 16 (1-3/16) with a better ruler. With an even better ruler the length might be 29 to 24 (1-5/24). It stands to reason that you can get as close to the length as you want this way. The true length may be very close to 1-1/4, say 1-249,999/1,000,000, but it is still a ratio—in that case 1,249,999 to 1,000,000—of two natural numbers.

The true length is like a limit. But since each of the ratios that come closer and closer to the true length are ratios of natural numbers, you would assume that the true length must also be a ratio of natural numbers.

The Pythagoreans did not just base their arithmetic on these ideas. They also based their religion upon them. We know little of their religion today (except they were forbidden to eat beans and thought that people could be reincarnated as animals) but we do know that the natural numbers had a central place in it.

When you think of the Pythagoreans, you probably think of another famous theorem—the *Pythagorean theorem*, which states that the square of the longest side of a right triangle is equal to the sum of the squares of the other two sides. This theorem implies that the diagonal of a square 1 unit on a side is equal to $\sqrt{2}$ units.

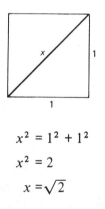

$$x^2 = 1^2 + 1^2$$
$$x^2 = 2$$
$$x = \sqrt{2}$$

So, following Pythagorean theories, there must be two natural numbers, call them a and b, such that the ratio of a to b, or a/b, is equal to $\sqrt{2}$.

Assume that a and b are in lowest terms. Since they have no common divisor, one of them must be even and the other odd, or they must both be odd. (If they were both even, 2 would be a common divisor.) Square both sides of

$$\frac{a}{b} = \sqrt{2}$$

to get

$$\frac{a^2}{b^2} = 2$$

Then you know that

$$a^2 = 2b^2$$

Since a and b are natural numbers, so are a^2 and b^2. And, since a^2 is twice a natural number, it is even. This in turn implies that a must be even, which is shown at the end of this proof. Since a is even, you can rewrite a as $2m$, where m is a natural number. Then you know that

$$\frac{2m}{b} = \sqrt{2}$$

Square both sides.

$$\frac{4m^2}{b^2} = 2$$

or

$$4m^2 = 2b^2$$

or

$$2m^2 = b^2$$

From the same line of argument used before, this equation implies that b must be even. But you know that a and b cannot both be even. Therefore, the original assumption that the square root of two could be expressed as the ratio of two natural numbers is false.

Here is an indirect proof that if a^2 is even, then a is even. Suppose a^2 is even and a is odd. Then a can be written as $2n - 1$, so $a^2 = (2n - 1)^2 = 4n^2 - 4n + 1$. Rewrite $4n^2 - 4n + 1$ as $2(2n^2 - 2n) + 1$. This is clearly odd, which contradicts the assumption that a^2 is even. Therefore, a must be even.

That $\sqrt{2}$ is not rational (*irrational*) is a remarkable result on several counts. Not only did it completely upset the foundation of the Pythagorean beliefs, but also it provided the first example of a theorem that could not be physically checked at all. If you prove that the base angle of an isosceles triangle are equal, then you can draw a few isosceles triangles, measure their base angles as closely as you can, and note that the theorem works. Even if you prove that something is not supposed to happen, say that two lines never meet, you can make careful drawings to check instances. But no drawing will ever show whether or not $\sqrt{2}$ is rational or irrational. For you can approximate $\sqrt{2}$ as closely as you like with ratios or rational numbers. For example, here are some increasingly good approximations: 14 to 10, 141 to 100, 1414 to 1000, and 14,142 to 10,000. Others would be even better. In other words, a theorem such as "the base angles of an isosceles triangle are equal" could be a theorem of physics. But the irrationality of $\sqrt{2}$ is a purely mathematical result.

Appalled by this result, the Greeks decided that lengths (or other magnitudes) ought to be treated separately from numbers. After the Pythagoreans, Eudoxus developed an excellent theory of "magnitudes" that were thought of as distinct from numbers. This gave Greek mathematics a split personality, emphasizing geometry at the expense of algebra. Who knows what Archimedes could have done with a workable algebra to back him up?

The Greeks may have had the right idea. When mathematicians finally decided how to incorporate irrational numbers into arithmetic, all kinds of problems resulted. (These are the subjects of Chapter 5.) It is still true that there is no direct physical application of irrational numbers, for no one can measure anything to a degree of precision greater than can be expressed by rational numbers. In a sense, irrational numbers are of theoretical interest only (although many parts of mathematics rely on irrational numbers, rational approximations must be used for calculations). Their sudden appearance in mathematics was not only a surprise, but, given what the Pythagoreans expected of mathematics, an unresolvable paradox. It is not considered a paradox today, because it does not contradict anything people know to be true. The Greeks, however, *knew* that you could measure any line segment to the precision you could see (and beyond) as the ratio of two natural numbers. That you could find line segments that could not be measured exactly in this way seemed impossible.

WHO SHAVES THIS PARADOX?

In mathematics what you take to be a paradox may be a proof, as you saw earlier. Or, what you see as a proof may lead to a paradox. Sometimes it is not easy to see which you are dealing with—paradox or proof.

Consider this 1918 proposition and question offered by Bertrand Russell. "A man of Seville is shaved by the Barber of Seville if and only if the man does not shave himself. Does the Barber of Seville shave himself?"

Clearly, if the answer is "yes," then the answer is also "no." And if the answer is "no," then the answer is also "yes."

CAN YOU FIND THE FLAW?

Hint: Is this paradox or proof?

Sometimes this is given in different formulations. For example, you might hear "in a certain village in Russia, every man shaves himself who is not shaved by the one and only barber in the village. Who shaves the barber?" One answer in this formulation is that she doesn't shave.

Another is that he has an enormous beard.

A somewhat less flip way to handle this is to say that this is a proof that no such village exists.

Step 1. Assume a village in Russia exists where every man shaves himself who is not shaved by the one and only (male) barber (who also is shaved).

Step 2. The barber cannot shave himself because every man who shaves himself is not shaved by the barber.

Step 3. Another person cannot shave the barber because every man who does not shave himself is shaved by the barber.

Step 4. Since the barber is shaved, the village does not exist.

When you look at the version involving Seville, this proof does not seem appropriate, for Seville certainly exists. Then the proof simply implies that the given situation cannot exist.

Here is another paradox that could be viewed in a similar way: Suppose a country declares that every city in the country must have a mayor. There are only two places that the mayor is allowed to live (and he or she must live in just *one* of them). Either the mayor must live in the city that he or she governs or the mayor must live in Mayor City, a special city set up just for mayors who do not live in the cities they govern. Since Mayor City is a city, it must have a mayor. Where should he or she live?

Another formulation of this same paradox involves the librarian who puts together a bibliography to be kept in his or her library of all those bibliographies in his or her library that do not list themselves. Should the new bibliography list itself?

It is easy to dispose of this kind of paradox by saying "Since this is self-contradictory, the situation described cannot exist." That is because this kind of paradox purports to speak of a real town or library. But what if a paradox describes a real situation—one that is known to exist but is self-contradictory? In that case, it might be that the law of the excluded middle does not always apply.

IT JUST FEELS TRUE

It is clear that in human experience no physical situation has been encountered where the law of the excluded middle fails. In mathematics, however, the situation is more complicated, for mathematics deals with situations that transcend physical experience.

For example, the irrationality of $\sqrt{2}$ is a phenomenon purely in mathematics. It has no counterpart in reality. Therefore, human physical experience with the real world cannot tell whether or not the law of the excluded middle should apply (although that law was used to establish the irrationality of $\sqrt{2}$). Furthermore, mathematics deals with infinities, which may not exist in the physical world. How do you know that something that happens only when infinities are used obeys the law of the excluded middle?

Some mathematicians, since 1907, have concluded that it is, at best, unsafe to apply the law of the excluded middle to infinities. These mathematicians, who claim that their intuition fails them at infinity, as known as *intuitionists*. They work only with entities that they can actually construct. Most existence or nonexistence proofs are ruled out. Irrational numbers, even, are highly suspect. Instead, they give examples of numbers that, by their rules, have no definite values. Since the "value of a number" *is* the number, a "number with no definite value" simply fails to exist.

Consider any unproved or unknown result in mathematics—for example, all of the digits of the decimal expansion of π. While computers have been used to calculate the decimal expansion in π to many, many places, it has been shown that π will have an infinite expansion and, until it is calculated, you cannot predict what the next digit will be. You can use this unknown part of π to define a number.

Say that you examine the most extensive calculation known of the expansion of π and you notice that there is no place that the digit 0 occurs more than four or five times in a row. Now define the number A as follows: A is represented as $0.000 \cdots$ but the nth digit of A is 1 if the nth digit of the decimal expansion of π is the first of ten 0s in a row; otherwise the nth digit is 0.

Now, it is now known whether or not there is a part of the decimal expansion of π that has ten 0s in a row. A computer might churn out digits forever (and not reach the end) without finding such a part of the expansion. If there is no such part, then $A = 0$. If there is such a part, then A is greater than 0.

By ordinary mathematics, A exists (although you do not know its value). By intuitistic mathematics, A does not exist. Even if the computer found one part of the expansion of π with ten 0's in a row, so that A could be shown to be greater than 0, you would still not know the value of A, for there might be another ten 0's in a row somewhere down the road.

But it is not necessary to appeal to the infinite to obtain intuitionistic results that appear paradoxical. Consider any unknown result—say *Goldbach's conjecture.* In 1742, Christian Goldbach (1712–1764) suggested to Euler that every even natural number greater than 2 is the sum of exactly two primes. For example, $4 = 2 + 2$, $6 = 3 + 3$, $8 = 3 + 5$, $10 = 5 + 5$, $12 = 7 + 5$, $14 = 11 + 3$, $16 = 11 + 5$, $18 = 11 + 7$, $20 = 13 + 7$, and so forth. As of this writing, no one knows whether or not Goldbach was right (although no one has been able to show him wrong). By ordinary mathematics, however, you can use Goldbach's conjecture to define a natural number: The number G is

137 if Goldbach's conjecture is true and G is 43 if G is false. You know that G exists, even though you do not know what it is. However, intuitionist mathematicians would say that G does not exist, since you cannot construct it.

None of this would seem odd to a physicist specializing in *quantum theory*, which is a branch of physics that deals with subatomic particles. According to quantum theory, which had its start around the same time as intuitionist mathematics, a phenomenon does not exist until it is observed. For example, if a single "particle" of light—called a *photon*—is aimed at a detector screen, you cannot tell where it will strike the screen until it is observed striking the screen. Before it strikes the screen, there are only probabilities, all less than 1, as to where it will strike. After the photon has been observed striking the screen, the probability that it will strike in that particular place becomes 1, while all the other probabilities become 0.

Furthermore, these probabilities can change in paradoxical ways. You can arrange the following experiment. Project the photon through one of two slits of a specified size and you can observe that there is an area in which the photon never strikes the screen—in other words, an area on the screen where the probability of observing the photon is 0. Then cover the slit you were not using but continue to project the photon through the same slit as before. Now the photon begins to land sometimes in the previously forbidden area. How does the photon "know" that you have covered the slit?

There is no really satisfactory answer to that last question—and a full discussion would involve you in a lot of physics. But notice how this result parallels the definition of G. You do not know whether the photon strikes the screen at A or B until you see it land. You do not know whether G is 137 or 43 until you prove or disprove Goldbach's conjecture. The quantum physicist says that there is no way to know where the photon is before it is seen. The intuitionist says that there is no way to discuss G until you have shown a way to construct it by proving or disproving the conjecture. Although most physicists believe quantum theory is true, however, most mathematicians are not interested in intuitionism. Perhaps mathematicians would feel differently if they knew a little more physics.

Erwin Schrödinger, the Nobel Prize-winning physicist, points out that if a particle is observed in one place and then a short time later a similar particle (in all respects) is observed nearby, it is natural to assume that you are observing two *instances* of the same particle. You can reason that the first particle must have come from someplace, so it must have a path. Also, the second particle had to be somewhere while the first observation was made, so it too has a path. But, like an intuitionist, Schrödinger does not allow you to deduce that it is *not* one particle on one path.

The intuitionist point of view seems to rule out much of ordinary mathematics. For example, in the discussion of

$$1 - \frac{1}{2} + \frac{1}{3} - \frac{1}{4} + \cdots + \frac{1}{n_{\text{odd}}} - \frac{1}{n_{\text{even}}} + \cdots$$

in Chapter 2, you were asked to accept that this series has a sum on the following basis:

> Since the sum of n terms of the series can never be greater than 1 and since it can never be less than $1/2$, there must be a sum for the infinite series.

However, this is a special case of an existence theorem that the intuitionists would not accept. This theorem, and others like it, form much of the basis for calculus.

REVERSE LOGIC

It is possible that the intuitionist point of view seems peculiarly fussy to you. Consider, however, how unrestricted use of logic can lead to *Hempel's paradox*. This paradox, first proposed by Carl G. Hempel in 1937, touches on the scientific method, probability theory, and—from the point of view of some intuitionists—formal logic.

Usually, Hempel's paradox is given in a scientific context. Say, for example, that you are an ornithologist investigating crows. The more you study the subject, the more often you observe that all of the crows in your study are black. You therefore state a hypothesis: all crows are black. To increase the probability that your hypothesis is true, you continue your field work, observing more and more crows, always noting that they are black. Each observation, or confirming instance, increases the probability that your hypothesis is correct.

Hempel asserted, however, that you need not go out in the field at all. Instead, you could sit in your office at home and make observations that would just as well increase the probability that your hypothesis is true.

To do this, you would make use of the *law of the contrapositive*, which in one form states that "All X is Y" is logically equivalent to "All not-Y is not-X." Another form of the law is that "If X, then Y" is equivalent to "If not-Y, then not-X." In either case, an observation that something not black is not a crow would tend to confirm the idea that something that is a crow is black.

Thus, sitting in your office, you may observe that the rug is red. A rug is not a crow, and red is not black. Therefore, you have observed that a not black thing is not a crow, which is the logical equivalent to observing that a crow is a black thing.

However, observing a red rug would equally be a confirming instance of "All crows are white."

Although the set of crows is finite, it is very large. For a small finite set, this method of reasoning presents no problems. For example, suppose that the original hypothesis had been "All pens on my desk are ball-points." To check this, I proceed to note that the things on my desk that are not ball-points are not pens. For example, the pencils are not ball-points and they are also not pens. If I check every not-ball-point on the desk, finding each to be a not-pen, however, I have established the truth of the original hypothesis. That is, had I found a not-ball-point that *was* a pen, the original proposition would be false. Not finding a not-ball-point that is a pen confirms the hypothesis—provided everything on the desk has been checked.

The problem with Hempel's paradox is that it deals with increased probability from confirming instances. It assumes that if I find one ball-point pen, the probability is greater that all pens will be ball-point. From that, Hempel deduces that if I find one non-ball-point pencil on the desk, it also increases the probability that all pens will be ball-point. Viewed this way, for finite sets, Hempel's paradox is a difficulty for probability theory, not for logic.

In mathematics, however, sets are often infinite (at least in the sense that you can always find one more member). Consider a statement such as Goldbach's conjecture: every even number greater than 2 is the sum of exactly two primes. No mathematician would admit that finding more and more instances of Goldbach's conjecture true would prove the result, although some might say that these only increase the probability of it being true. However, most mathematicians, including some, but not all, intuitionists, would agree that the law of the contrapositive holds. Applied to Goldbach's conjecture, this gives "All things that are not the sum of exactly two primes are not even numbers greater than 2." Therefore, since 2 + 4 = 6, and 4 is not a prime, but 6 is an even number greater than 2, Goldbach's conjecture is false.

CAN YOU FIND THE FLAW?
Hint: State Goldbach's conjecture
in a more exact form.

A more exact statement of Goldbach's conjecture in mathematical language might be: "If n is an even number greater than 2, then there exist prime numbers a and b such that $a + b = n$." The contrapositive of that statement is "If there do not exist prime numbers a and b such that $a + b = n$, then n is not an even number greater than 2." In that form, 2 + 4 = 6 is not a refutation of Goldbach's conjecture.

You can simplify the statement of the contrapositive without changing its meaning by saying: "If, for a number n, there do not exist prime numbers a and b such that $a + b = n$, then n is an odd number greater than 1." In this form, it is easy to find confirming instances of this. Since all odd numbers must be the sum of an even and an odd number, and since 2 is the only even prime, then every prime n that is not the greater of pair of twin primes is a confirming instance. (Twin primes are primes such as 3 and 5, 29 and 31, or 57 and 59 which are separated by 2 exactly. It is not known whether or not there is an infinite number of them.) There are an infinite number of primes that are not twin primes. Thus, since there are an infinite number of confirming instances of Goldbach's conjecture (in this form), it must be true.

This somewhat tortured illustration suggests that the implications of Hempel's paradox do not apply to mathematics. Finding an infinite number of confirming instances of the contrapositive would not convince any mathematician of the truth of the original statement. Most mathematicians would not abandon the contrapositive, however; even most intuitionists would not. It is the notion of "confirming instance" as increasing the probability of truth that does not apply to the infinite.

Even in its original form, you can easily find an infinite number of confirming instances of Goldbach's conjecture. There are an infinite number of odd primes. The sum of any two of them is an even number. But that observation does not establish Goldbach's conjecture as true, for somewhere in the infinity of even numbers there may be one that is not the sum of two primes.

4
Speaking with Forked Tongue

Gentlemen, that $(e^i + 1 = 0)$ is surely true, but it is absolutely paradoxical; we cannot understand it, and we don't know what it means, but we have proved it, and therefore, we know it must be the truth.

Benjamin Pierce

You are now going to abandon mathematics, strictly speaking, for a chapter. This is not just for a rest. Many people view mathematics as a language, so it is helpful to look for a time at the workings (or, in this case, nonworkings) of language. Additionally, the problems that arise from language carry over into mathematics in quite a direct way—but they are easier to see clearly in a "natural" language as opposed to the "artificial" language of mathematics.

A COLLECTION OF CONTRADICTIONS

The only way to get rid of temptation is to yield to it.	Oscar Wilde
We have met the enemy, and he is us.	Walt Kelley
Please ignore this notice.	Graffiti
All Cretans are liars.	Epimenides the Cretan
The statement I am making is false.	Eublides

To which one might add, "Damned if I do, and damned if I don't," for all of these contradictory remarks not only lead you down the garden path, but they lead you down two paths at the same time. There are, however, different degrees of self-contradiction involved.

Oscar Wilde's epigram, "The only way to get rid of temptation is to yield to it," for example, appears to be a substantially true statement. Certainly, one way to get rid of temptation is to yield. For if you yield to it, the temptation no longer exists. (Wilde was pushing it, however, when he called yielding the *only* way. For example, you can kill yourself.)

Walt Kelley's remark, "We have met the enemy, and he is us," put into the mouth of the comic-strip character Pogo, became a rallying cry of the environmental movement. Like Wilde's epigram, it was certainly taken by those who were marching against pollution (or for population control) as substantially true. There is a bit more interpretation required, however. The paradoxical sound to the slogan comes from the same source as the somewhat older remark, "He is his own worst enemy." Both sentences can be rephrased in ways that are meaningful without having that contradictory quality. The environmentalists meant "We, who are the victims of pollution, are also the cause of pollution." "He is his own worst enemy" is another way of saying, "As he goes about his business, he foolishly creates situations that cause trouble for himself."

The implicit contradiction in Kelley's formulation can be seen more easily by noting that an enemy of someone is a force that is trying to do the someone harm. That force is generally outside the someone. While it seems paradoxical for it to be inside the someone (or the same as the someone), there is nothing inherent in the language that prevents this.

When you turn to the next three quotations, however, the situation becomes more precarious. Of course, "Please ignore this notice" is not a statement, but a command. Since only statements can be true or false, there is no question as to the truth or falsity of "Please ignore this notice." It is moot. However, how can you obey the command? If you ignore the notice, you do what it tells you to do, so you are not ignoring it. If you do not ignore the notice, then you have to pay some attention to it.

A more extreme example of the "Please ignore this notice" type (as well as a less polite one) is as follows. You walk into a room full of books. On the wall, there is a sign that says "Do not read anything in this room." To do what the sign says, you would have not to read the sign. But if you do not read the sign, then you will not be able to do what it commands. Yet, as far as the situation itself goes, such a sign seems perfectly reasonable. In fact, if you replace the sign by a librarian who tells you "Do not read anything in this room" as you enter it, there is no paradox at all. The paradoxical quality comes from the source of the sentence rather than from the sentence itself. Similarly,

is a nonparadoxical way to cancel an announcement. Only when there is no other notice around does "Please ignore this notice" cause problems.

Similarly, if Epimenides had not been a Cretan, there would be no problem with the truth or falsity of "All Cretans are liars." However, since Epimenides was a Cretan, his statement must be a lie. But if it is a lie, then not all Cretans are liars, so the statement could be true.

St. Paul was not a person to be bothered by such complexities. In his letter to Titus, obviously referring to the Epimenides paradox, he noted, "One of themselves, a prophet of their own, said, 'Cretans are always liars, evil beasts, lazy gluttons.' This testimony is true." Of course, Paul was just being carried away by his own poor opinion of Cretans. He winds up this passage by saying of the Cretans that "they are detestable, disobedient, unfit for any good deed."

The granddaddy of all paradoxes is the Eublides statement, "The statement I am making is false." There are no outside circumstances. In fact, in the original form, it may simply have been "I am lying." If this is a lie, then it is the truth. If if is a truth, then it is a lie. Eublides may also be the author of the Epimenides paradox, in which case he has the record for the oldest of the common paradoxes, since he lived in the fourth century B.C.

THE LIAR

Since then, everyone, including of course Aristotle, has discussed one form or another of Eublides paradox, which the Greeks called *The Liar*. In Aristotle's version, it was "This statement is false."

By the Middle Ages, philosophers had developed a dialogue version of The Liar:

Socrates: What Plato is about to say is false.
Plato: Socrates has just spoken truly.

In 1913, P.E.B. Jourdain created a modern version of the dialogue, this time without any speakers. On one side of a card is printed

"The statement on the other side of this card is true," while on the other side it says, "The statement on the other side of this card is false."

Alfred Tarski developed a version that resembles mathematical induction. You have a 100-page book. On page 1 it says, "The sentence on page 2 of this book is true." On page 2 it says, "The sentence on page 3 of this book is true." In general, on page n it says, "The sentence on page $n + 1$ of this book is true" *except for page 100*, where it says, "The sentence on page 1 of this book is false." Basically, Tarski's version is the same as Jourdain's. He is just keeping you in suspense a while longer.

José Benardete, with another purpose in mind, created a book with an infinite number of pages (see Chapter 8). Suppose that Tarski's version of The Liar were adapted to fit that book. On the first page, you could write "The sentence on the last page of this book is true." On every other page, you could write "The sentence on the page before this is false." When you read page 1, there is no way of checking the truth or falsity of the statement, for there is no last page of the book. But when you read page 2, you discover that the sentence on page 1 was false. However, when you read page 3, you learn that page 2 was false, so page 1 must be true. But page 4 reverses the truth value of page 1 also.

The question, then, is "Is the sentence on page 1 true or false?"

There are various explanations of these paradoxes. None of them is completely satisfactory, which is why people keep trying. Kurt Gödel, for example, gave this version of The Liar. On May 4, 1934 a person makes just one statement, which is "Every statement that I make on May 4, 1934, is false." Gödel then declared that this version of The Liar was a proof that "false statements in English" cannot be expressed in English. This "proof" is completely unconvincing, yet Gödel was the greatest logician since Artistotle.

DON'T CALL ATTENTION TO YOURSELF

Most discussions of The Liar, however, involve ruling out the kinds of statements that produce this paradox. These statements are defined as statements that apply to themselves. While this approach certainly eliminates "This statement is false," not everyone agrees that it works for all versions.

For example, in Jourdain's version, each sentence standing alone does not refer to itself. You have to take the two sentences together to obtain self-reference. Even then, you need the physical condition of a card with two sides. If you take one end of the card and give it a half twist, then glue one end carefully to the other, you will get a Möbius strip—an object with only one side. Then the sentences, instead of being paradoxical, become nonsense. But the sentences have not changed, only the reality in which they are embedded.

If St. Paul says "All Cretans are liars," there is no paradox.

In Tarski's book paradox, you need to consider 100 sentences together to find self-reference. In the infinite book version of The Liar, self-reference is obvious, but involves a countable infinity of statements. (It could be argued that it involves only two statements, one of which is repeated an infinite number of times. It is easy to modify this so that all of the statements are different by making each statement on page $n-n$ greater than 1—be of the form "The sentence on page $n - 1$ is false.")

It is difficult to define exactly what is meant by self-reference in any case. Henri Poincaré coined the technical term *impredicative* to describe the kind of self-reference that would not be allowed into mathematics. In 1912, Poincaré gave a murky definition of *impredicative definition*. He defined *impredicative definition* as a definition by a relation between the object being defined and *all* the individuals of the type that the object being defined is supposed to have itself as a part. This is not very clear, but it is generally taken to mean simply that whatever involves all of a collection must not be one of the collection (although Poincaré expressed the idea for definitions instead of for collections, or sets). While a general notion of self-reference is difficult to express, it is relatively easy to describe what is meant by self-reference for sets or other types of mathematical objects.

You can construct a definition of *self-reference* with The Liar specifically in mind. Here is one: "A sentence (or collection of sentences) is self-referential if the subject of the sentence is itself." You can then say that all self-referential sentences are to be assigned no truth value, which would eliminate The Liar paradox.

However, such a definition would also apply to the following, which seem at first glance to be either true or false in the normal way:

This sentence has five words.
This sentence has seven words.
This sentence is self-referential.
This sentence is in French.

Worse yet, the definition would not apply to sentences that say the same thing as the ones above:

Five words are used in this sentence.
Seven words are used in this sentence.

Of course, the truth values, if they exist, are reversed in the rewording, but it is no longer clear that the reworded sentences are self-referential. If "Five words are used in this sentence" is about "five words," what five words are they?

All this is complicated by the fact that some self-referential sentences are true in one language and false in another. For example, "This sentence is in French" is true when translated into French, but false as it stands. A sentence with such a wavering truth value ought not be used in mathematics.

Therefore, while it would be pleasant to allow the sentence "This sentence has five words" be accepted as true, it is definitely safer to say that the truth values of self-referential sentences are *relative*. In all cases (if you include the collections of sentences that are self-referential), the truth or falsity of a self-referential sentence depends on the circumstances.

VICIOUS, VICIOUS, VICIOUS

Here is a circumstance that illustrates how difficult it is to apply this criterion. There are many versions of this paradox. You may recall the following version from Cervantes' *Don Quixote*.

Sancho Panza has become governor of the island of Barataria, where it is the law that every person coming to a part of the island must state his or her reason for coming. If the person tells the truth, he or she must go free. If the person lies, however, he or she will be hanged. A traveler arrives and announces his intention as "I am here to be hanged." Now Sancho has a problem. If the man is telling the truth, he should go free; but it would not be the truth unless he were hanged. On the other hand, if the man is lying, he should be hanged — in which case his statement was not a lie.

There are, as noted, other versions of the same idea. One that is particularly ancient involves a crocodile that has captured a child. The crocodile offers to return the child to its father if the father can guess whether or not the child will be returned. The father guesses that the crocodile will not return the child. Now the crocodile has got itself into a frightful logical mess.

The crocodile version has this advantage over Sancho Panza's problem: it does not directly involve the slippery notions of truth and falsity. It is, however, an artificial situation, rather like the Swedish civil-defense exercise paradox. That is, the crocodile's response can be *unexpected*, for it can change its reptilian mind when confronted with the father's reply.

Here is a version that does not involve quite such an artificial situation. Protagoras the lawyer, to show what a good teacher he is, contracts with his students that they must pay him for his instruction if and only if they win their first case. There is no fee if they lose. However, when one of his students has completed the course, he avoids taking any cases, and thus avoids payment. Protagoras decides his only recourse is to sue the student for the fee. When the case comes to court, the student represents himself. If he loses to Protagoras, then by their contract he does not have to pay. If however, he wins the suit, then the judgment will be that he does not have to pay.

Mathematicians usually classify these as more extensive versions of The Liar. Actually, they involve self-reference in a larger sense than The Liar does. This extended kind of self-reference is often called a *vicious circle*. An essential difference between this type of vicious circle and mere self-reference is that these paradoxes are not specifically paradoxes of language.

In fact, you can find vicious circles that have no language in them at all. *The Guinness Book of Records* includes a record for "the unsupported circle," a sporting event in which a large number of people sit on each others' knees in a circle. What is holding them up?

None of the events in the Protagoras paradox taken individually is unusual. In fact, the whole sequence would be possible in real life.

But what would happen in real life? Probably the judge, seeing that the student was evading payment, would order the student to pay. If he did not, contract notwithstanding, the student would have his property confiscated. This might be "unexpected" for the student, who thought he could get away with his scheme. But, like the Swedish civil-defense exercise, it would happen despite the student's logic.

In other words, the vicious circle paradoxes, while loosely related to The Liar, are really examples of situations in which reality goes one way while logic goes the other.

Nevertheless, these very artificial situations, which were invented primarily to produce paradoxes by pushing ordinary logic into absurdity, show vital connections between what is and is not possible in mathematics. The relationship of mathematics to reality is still being debated. When you dispose of a paradox by saying that it does not exist in reality, as in the Barber paradox, or by saying that in reality there would be no paradox, as in the Protagoras paradox, you do not necessarily eliminate similar paradoxes from mathematics. If they are entirely within mathematics, appeals to reality do not help.

AN OBSCENE PARADOX

With this in mind, it is useful to look at some paradoxes that are essentially in language, but which can be (and will be in later chapters) mapped into mathematics.

The first is a version of The Liar invented by Willard Van Orman Quine. There is a charming discussion of Quine's paradox in Douglas Hofstadter's book *Gödel, Escher, Bach: an Eternal Golden Braid*. It is one of his dialogues between Achilles and the Tortoise. Here is an abridged version.

Achilles: You don't mind if I change the subject, do you?

Tortoise: Be my guest.

Achilles: Very well, then. It concerns an obscene phone call I received a few days ago.

Tortoise: Sounds interesting.

Achilles: Yes. Well—the problem was that the caller was incoherent, at least as far as I could tell. He shouted something over the line and then hung up—or rather, now that I think of it, he shouted something, shouted it again, and hung up.

Tortoise: Did you catch what that thing was?

Achilles: Well, the whole call went like this:

>*Myself:* Hello?
>
>*Caller (shouting wildly):* Yields falsehood when preceded by its quotation! Yields falsehood when preceded by its quotation!
>
>(*Click.*)

Tortoise: That is a most unusual thing to say to somebody on an obscene phone call.

Achilles: Exactly how it struck me.

Tortoise: Perhaps there was some meaning to that seeming madness.

Achilles: Perhaps . . . .

Tortoise: I've been thinking about that obscene phone call. I think I understand it a little better now.

Achilles: You do? Would you tell me about it?

Tortoise: Gladly. Do you perchance feel, as I do, that that phrase "preceded by its quotation" has a slightly haunting quality about it?

Achilles: Slightly, yes—extremely slightly.

Tortoise: Can you imagine something preceded by its quotation?

Achilles: I guess I can conjure up an image of Chairman Mao walking into a banquet room in which there already hangs a large banner with some of his own writing on it. Here would be Chairman Mao, preceded by his quotation.

Tortoise: A most imaginative example. But suppose we restrict the word "preceded" to the idea of precedence on a printed sheet, rather than elaborate entries into a banquet room.

Achilles: All right. But what exactly do you mean by "quotation" here?

Tortoise: When you discuss a word or a phrase, you conventionally put it in quotes. For example, I can say,

The word "philosopher" has five letters. Here, I put "philosopher" in quotes to show that I am speaking about the WORD "philosopher" rather than about a philosopher in the flesh. This is called the USE-MEN-TION distinction.

Achilles: Oh?

Tortoise: Let me explain. Suppose I were to say to you, Philosophers make lots of money.

Here I would be USING the word to manufacture an image in your mind of a twinkle-eyed sage with bulging money-bags. But when I put this word—or any word—in quotes, I subtract out its meaning and connotations, and am left only with some marks on paper, or some sounds. That is called "MENTION." Nothing about the word matters, other than its typographical aspects—any meaning it might have is ignored . . . . But now, I want you to think about preceding something by its own quotation.

Achilles: All right. Would this be correct?
"HUBBA" HUBBA

Tortoise: Good. Try another.

Achilles: All right.
"'PLOP' IS NOT THE TITLE OF ANY BOOK, SO FAR AS I KNOW"
'PLOP' IS NOT THE TITLE OF ANY BOOK, SO FAR AS I KNOW.

Tortoise: Now this example can be modified into quite an interesting specimen, simply by dropping 'Plop.'

Achilles: Really? Let me see what you mean. It becomes
"IS NOT THE TITLE OF ANY BOOK, SO FAR AS I KNOW"
IS NOT THE TITLE OF ANY BOOK, SO FAR AS I KNOW.

Tortoise: You see, you have made a sentence.

Achilles: So I have. It is a sentence about the phrase "is not the title of any book, so far as I know," and quite a silly one, too . . . .

Tortoise: Not to my mind. It's very earnest stuff, in my opinion. In fact this operation of preceding some phrase by its quotation is so overwhelmingly important that I think I'll give it a name.

Achilles: You will? What name will you dignify that silly operation by?

Tortoise: I believe I'll call it "to quine a phrase," to quine a phrase.

Achilles: "Quine"? What sort of word is that?

Tortoise: A five-letter word, if I'm not in error . . . .

Achilles: Anyway, now I know how to quine a phrase. It's quite amusing. Here's a quined phrase:

"IS A SENTENCE FRAGMENT" IS A SENTENCE FRAGMENT.

It's silly, but all the same I enjoy it. You take a sentence fragment, quine it, and lo and behold, you've made a sentence! A true sentence, in this case.

Tortoise: How about quining the phrase . . .

"WHEN QUINED, YIELDS A TORTOISE'S LOVE SONG"

Achilles: That should be easy . . . I'd say the quining gives this:

"WHEN QUINED, YIELDS A TORTOISE'S LOVE SONG"
WHEN QUINED, YIELDS A TORTOISE'S LOVE SONG.

Hmmm . . . There's something just a little peculiar here. Oh, I see what it is! The sentence is talking about itself! Do you see that?

Tortoise: What do you mean? Sentences can't talk.

Achilles: No, but they REFER to things—and this one refers directly—unambiguously—unmistakably—to the very sentence which it is! You just have to think back and remember what quining is all about.

Tortoise: I don't see it saying anything about itself. Where does it say "me," or "this sentence," or the like?

Achilles: Oh, you are being deliberately thick-skulled. The beauty of it lies in just that: it talks about itself without having to come right out and say so! . . .

Tortoise: Now tell me: is the following sentence self-referential?

"IS COMPOSED OF FIVE WORDS" IS COMPOSED OF FIVE WORDS.

Achilles: Hmmm . . . I can't quite tell. The sentence which you just gave me is not really about itself, but rather about the phrase "is composed of five words." Though, of course, that phrase is PART of the sentence . . .

Tortoise: So the sentence refers to some part of itself—so what?

Achilles: Well, wouldn't that qualify as self-reference, too?

Tortoise: In my opinion, that is still a far cry from true self-reference. . . . But for now, why don't you try quining the phrase "yields falsehood when preceded by its quotation?"

Achilles: I see what you're getting at—that old obscene phone call. Quining it produces the following:

"YIELDS FALSEHOOD WHEN PRECEDED BY ITS QUOTATION"
YIELDS FALSEHOOD WHEN PRECEDED BY ITS QUOTATION.

So this is what the caller was saying. I just couldn't make out where the quotation marks were as he spoke. That certainly is an obscene remark! People ought to be jailed for saying things like that!

Tortoise: Why in the world?

Achilles: It just makes me very uneasy. Unlike the earlier examples, I can't quite make out if it is a truth or a falsehood. And the more I think about it, the more I can't unravel it. It makes my head spin. I wonder what kind of a lunatic mind would make something like that up, and torment innocent people in the night with it?

Tortoise: I wonder . . .

As the dialogue suggests, Quine devised an artificial sentence that, technically speaking, is not self-referential but still partakes of the essence of The Liar. Actually, in Quine's version, the sentence is the following: "Yields a false conclusion when appended to its own quotation" yields a false conclusion when appended to its own quotation. Hofstadter's version makes a better shout over a telephone, however.

It would appear that any method of ruling out Quine's paradox would also rule out perfectly normal sentences in English. For example, the sentence "'Quine' has five letters" (an acceptable response to the question, asked by someone doing a crossword, "What's a five-letter word for 'Harvard philosopher'?") is the same construction as Quine's paradox. Some of Hofstadter's examples, which involve quining, ought to be allowable in English: for instance, "Is a sentence fragment" is a sentence fragment.

Furthermore, the Quine paradox is not dependent on the language in which it is expressed or its exact wording. In particular, the Quine paradox can be translated into a statement in mathematics. However, the odd thing that occurs in that translation (if it is carefully done) is that the result is no longer a paradox. You will see how this works in a later chapter.

In Chapter 1, you saw how a contradiction can be eliminated by agreeing that some operations (in that case, division by zero) are not allowable. You also saw that definitions can be carefully made or modified (for complex numbers) to avoid contradictions. In Chapter 3, you saw that the existence of a contradiction that cannot be eliminated, such as the assumption of a rational square root for 2, is taken by most mathematicians to mean that the opposite of the statement that leads to the contradiction must be true. None of these approaches seems to have any effect on Quine's paradox. The contradiction is within itself.

THE WORD TURNS

Here is the second paradox about language that can be carried over into mathematics. This one was invented by Kurt Grelling in 1908.

Begin by noting that some adjectives describe themselves and others do not. For example, *short* is a short word and *polysyllabic* is polysyllabic, but *red* is not red and *long* is not long. Now classify the adjectives that describe themselves as *autological* (which means "self-descriptive") and those that do not describe themselves as *heterological* (or "non-self-descriptive"). This classification should work for all adjectives. In fact, most adjectives turn out to be heterological.

Adjective	Classification
monosyllabic	heterological
inexpensive	heterological

English	autological
French	heterological
unlikely	heterological
autological	autological
heterological	?

If heterological describes itself, then it is autological, so it does not describe itself. If heterological does not describe itself, then it is heterological, so it does describe itself.

Heterological is a good example of the kind of thing Poincaré was trying to avoid by ruling out impredicative definitions. The set of adjectives was divided into two parts by a particular property. Then that property was treated as an adjective to find out which of the two parts contained it. Poincaré said (probably correctly) that if you refrain from doing that, then Grelling's paradox—and similar paradoxes in mathematics—cannot arise. Of course, Poincaré would also rule out asking whether or not *autological* is autological, even though that does not seem to cause problems. Furthermore, when Poincaré's rules are strictly applied to mathematics, very many seemingly legitimate definitions must be eliminated. While there exist such things as undefined terms in mathematics, it would not make sense to leave these definitions undefined. For example, the definitions to be eliminated would include many of the essential ideas of the calculus, such as the least upper bound, which would be hard to do without.

Also, strict applications of Poincaré's rule to language would seem to eliminate some perfectly reasonable questions. For example, some words are onomatopoeic, such as *plop*, *buzz*, or *zoom*. It ought to be possible to ask whether or not the word *onomatopoeic* is onomatopoeic or not.

DOES PEGASUS EXIST?

Although this chapter is concerned with paradoxes that arise in language, there is one paradox that belongs in this group that involves both language and mathematics. In fact, the paradox is sometimes known as the Word Paradox. It is based on a 1905 paradox invented by Jules Richard, but in this chapter you will be concerned with the

1908 version invented by G.G. Berry. *Richard's paradox* involves a particular kind of complicated argument that is discussed in the next chapter. After you have seen that kind of argument in use, Richard's paradox will be presented.

The original version of Berry's paradox goes like this: You know that numbers can be referred to either in Hindu-Arabic numerals (for example, 2, 33, or 1,456) or in English words (for example, two, thirty-three, or one thousand four hundred fifty-six). Now it is apparent that the number of syllables in the English names tend to increase as the numbers become greater. In fact, for any given number of syllables, you can find a name that has more syllables. In particular, there must be names with more than nineteen syllables.

A well-known property of the natural numbers (which are also called the positive integers) is that every set of natural numbers has a least member (but not always a greatest member). Hence, *the least integer not nameable in fewer than nineteen syllables* must name a definite number. In fact, using the ordinary method of naming numbers it is 111,777. However, *the least integer not nameable in fewer than nineteen syllables* has only eighteen syllables. So, the least integer not nameable in fewer than nineteen syllables can be named in eighteen syllables.

As originally formulated, Berry's paradox is startling, but sloppy. It seems as if the rules for naming numbers are changed part way through the problem. Hence, various people have reformulated it slightly to get rid of these difficulties. Here is one reformulation.

To start out, agree in the beginning that any way of naming the number in English will be acceptable. For example, admit *two*, *the successor of one*, *the only even prime*, *the first even natural number*, and so forth, on an equal basis. Then divide the natural numbers into two sets: those that have a name with fewer than 100 letters and those whose shortest name has more than 100 letters. (For simplicity, count the space between words as a letter.) The first set has to be finite, since there are only 26 letters in English and the space, so you have 27 choices for the first letter, 27 choices for the second letter, and so forth. In fact, this shows that the number of names will be less than 27^{100}. Since there is a countable infinity of natural numbers, there must be numbers that cannot be named in fewer than 100 letters. Now consider *the least number that cannot be named in a hundred letters or fewer.* But that number has just been named in 67 letters (counting the spaces as letters).

You might also want to note that if you use a mechanical process for laying out the 27^{100} names, you would find *the least number that cannot be named in a hundred letters or fewer* in the list. Say that you started with 100 spaces as your first possible name. Then the second might be 99 spaces followed by *a*. The third would be 99 spaces followed by *b*. The list would eventually include the paradoxical name with 67 letters in it.

Logicians argue with each other from time to time about such questions as whether or not having a name indicates some sort of existence. For some reason, they usually give Pegasus, the winged horse of Greek myth, as an example. Since Pegasus, the constellation of stars, certainly exists (as much as anything can be shown to exist), it might be a better choice of argue whether or not the least number that cannot be named in a hundred letters or fewer exists. Like Pegasus, it has a name. For most nonlogicians, possession of a name does not guarantee existence. In fact, it is common to give names to things that do not exist. Berry's paradox names a number that does not exist.

On the other hand, the same reasoning could not be used to say that the square root of two does not exist. It would be true that the square root of two would not exist if all numbers were ratios of rational numbers. But it seems better to change the definition of number than to declare nonexistence for the square root of two. After all, you can visualize the square root of two in various ways. Not only can you not visualize the least number not nameable in a hundred letters or fewer, except to think that it should be fairly large, but also there is no convenient assumption that you can give up. The only assumption is that all numbers can be named in English, which is certainly true.

When you look at Richard's original version, upon which Berry's paradox is based, the question of existence becomes more pressing.

5
Paradoxes That Count

If we do run into a paradox, we can probably save the structure of mathematics by patching it.

Andrew Gleason

The classic Greek mathematicians were as suspicious of infinity as they were of irrational numbers. Aristotle declared that there was no such thing as a completed infinity. Instead, there was only a possibility of extending a finite set or a finite figure. In his book *Geomtry*, Euclid took great care to state what you know as the Fifth, or Parallel, Postulate in such a way that he did not need to consider the infinite. Instead of using something like "through a point not on a line there is one and only one parallel to the line," he postulated conditions under which two lines would meet. Even then, he was obviously doubtful about the postulate and avoided it in proofs where he could. For example, Euclid's famous proof about primes is usually taken to be a proof that the number of primes is infinite. Actually, Euclid avoided the infinite by showing there is no *last* prime.

Later mathematicians expressed the same skepticism, notably Gauss (although Euler was unafraid of the completed infinity and made a number of errors as a result). When, at the end of the nineteenth century, mathematicians began to explore completed infinities, those who worked with infinity were attacked in vicious terms by those who did not believe in the existence of completed infinities.

(It is the infinitely large that was causing trouble in the late nineteenth and early twentieth centuries. Since the time of Newton and Leibniz, mathematicians had used the infinitely small with great results, but considerable criticism. Mathematicians did not have a rigorous explanation of why their methods worked until the middle of the nineteenth century, at which point explanations were introduced that eliminated the concept of the "completed infinitely small," or infinitesimal, from formal mathematics. The infinitesimal continued to be used informally for many years. Finally, the idea was dropped. At that point, in the twentieth century, Abraham Robinson reintroduced the infinitesimal to mathematics on a new basis. Most mathematicians, however, prefer to use methods that avoid infinitesimals.)

It is not surprising that mathematicians rejected the notion of a completed infinity. First of all, most models of the universe assumed that it was finite. Except for God, who was outside the universe in some sense (as its creator, for instance), how could a completed infinity exist in a finite universe? This argument was particularly telling with regard to such ideas as a line of infinite length. In fact, Gauss's objections to infinity were principally to infinite magnitudes.

Philosophers also had trouble with the notion of a completed infinity of natural numbers. If the natural numbers were abstractions of physical properties (just as the color red is an abstraction of a characteristic of red objects), then where was the natural object (or set of natural objects) from which infinity could be abstracted? If, on the other hand, numbers were constructs of the human mind, how could a finite mind construct a completed infinity?

An available model for a completed infinity was the number of points on a line. Zeno of Elea, however, had offered many arguments that raised havoc with this notion. (Zeno's arguments get a chapter to themselves later.)

Finally, it must have been clear from early times that infinity produced paradoxes beyond those we know from Zeno (although many of Zeno's paradoxes have been lost). Although these paradoxes are usually attributed to Galileo, it is difficult to believe that the Greeks, with their skepticism about completed infinities, were unaware of them.

A GOOD COUNT

There are two basic methods of counting in common use throughout the world. One method is what is usually meant when you say that a child has "learned to count." This generally means that the child has developed a set of names for numbers that are arranged in the order of the natural numbers: one, two, three, and so forth. To count a set of objects, the child matches each object with a name for a number in order. When the child reaches the last name, that is the name for the number of objects. The objects have been counted.

The other method, which probably is older, involves directly matching two sets of objects. It is believed that shepherds, for example, kept a bag of stones that could be matched with the number of their sheep. If there were stones left over when the stones and sheep had been matched, then some sheep were missing (obviously the same number of sheep as the leftover stones). Before writing was invented, traders sent hollow baked clay balls along with their merchandise. When the shipment arrived, the receiver could break open the ball and match the objects within the ball with the goods that had been received. If the correspondence between the objects from the sealed ball and the goods received was exact, then the receiver knew that nothing had been stolen along the way.

This method of comparing numbers without actually counting them in the ordinary way is known as *one-to-one correspondence*. For finite sets of objects, it is clear that when there is a one-to-one correspondence between them, they have the same number. In fact, that is the way number is defined for children in first grade and for working mathematicians. (Mathematicians who are involved with the foundations of mathematics start with undefined terms and mathematical induction.) When applied to infinite sets, however, this method results in paradoxes.

Consider the countable infinity of natural numbers, for example. Intuitively, you know that there are as many even numbers as there are odd numbers. Indeed, it is easy to match them in a one-to-one correspondence.

$$2 \quad 4 \quad 6 \quad 8 \quad 10 \quad \cdots \quad 2n \quad \cdots$$
$$\updownarrow \quad \updownarrow \quad \updownarrow \quad \updownarrow \quad \updownarrow \qquad\qquad \updownarrow$$
$$1 \quad 3 \quad 5 \quad 7 \quad 9 \quad \cdots \quad 2n-1 \quad \cdots$$

For every even number, there is an odd number. For every odd number, there is an even number.

There must be more natural numbers than perfect squares, however. (This is the example given by Galileo in 1638.) As you get to greater and greater numbers, the perfect squares become farther and farther apart. But the natural numbers can be matched one-to-one with the perfect squares.

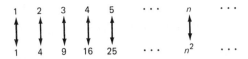

In fact, it turns out that all the infinite sets of natural numbers match one-to-one with each other. Perfect cubes, even numbers, primes (although the correspondence cannot be shown), and so forth. Intuition tells you that there ought to be twice as many natural numbers as there are even numbers, but one-to-one correspondence tells you that there are just as many numbers in each set.

Notice that this relationship is not true for finite sets. Instead, finite sets fulfill your expectations exactly. There are 50 even numbers less than or equal to 100 and there are twice as many natural numbers less than or equal to 100.

Since infinite sets lead to paradoxes when they are matched by one-to-one correspondence, one solution to the paradoxes is to declare that only finite sets can exist.

It is not just the numbers that cause problems, either. Consider the points on a line segment—that part of a line that is bounded by two endpoints. Surely there should be more points on a longer segment than on a shorter segment. If segment AB is twice as long as BC, then AB should have twice as many points.

Put the points in a one-to-one correspondence this way. "Bend" the segment AC sharply at B to form an angle.

Now connect A and C with a line. By the parallel postulate, there will be one and only one line through any point on AB that is parallel to line AC. You may recognize this procedure as related to a way of dividing a line segment into n parts. Say n is 100. If you divide AB into 100 equal parts, the parallels through those points will also divide BC into 100 equal parts. In fact, if you draw a parallel at each point of AB, each parallel intersects BC at only one point—and there are no points left over. The shorter line segment has as many points as the longer segment.

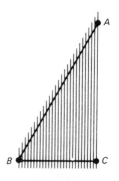

These paradoxes, as unsettling today as they were when first observed, are different in a fundamental way, however, from the fallacies of the first three chapters. The paradoxes of counting infinite sets by one-to-one correspondence do not involve contradictions within mathematics. These paradoxes establish that, on the basis of one-to-one correspondence, one infinite set, which intuition tells you is twice the size of another infinite set, has the same number of members as the other. The paradoxes do not come right out and say, for example, 2 = 1, as the paradoxes of Chapter 1 did. When mathematicians get a truly contradictory result, such as 2 = 1, they change the rules to avoid that kind of result. But a contradiction with intuitive ideas can be accepted, as you saw in the case of the irrationality of the square root of two.

PARADOX INCORPORATED

In the late nineteenth century, a mathematician named Georg Cantor (1845–1918), who was working with problems in the analysis of heat, decided to go against intuition. He would accept one-to-one correspondence as the means of comparing infinite sets. Furthermore, he would accept completed infinities into mathematics. This led eventually to the conclusion, expressed by Richard Dedekind, that the distinguishing characteristic of infinite sets was just this paradox. An infinite set is one in which, so to speak, the whole is not equal to the sum of the parts. More technically, it is one in which a part of the set that does not contain all the members of the set can be put into one-to-one correspondence with the whole set.

Here is another answer for the question: What to do with a paradox? If you are sure that no contradiction results, incorporate the paradox into mathematics and declare it a paradox no longer.

Cantor began to use one-to-one correspondence to compare infinite sets. You have already seen how various infinite parts, or *subsets*, of the natural numbers have the same quantity as the natural numbers. Cantor called this quantity the *cardinality* of the set. The set of even numbers, for example, has the same cardinality as the set of natural numbers. So does the set of perfect squares.

But what about the integers, both positive and negative? Do they have the same cardinality as the natural numbers?

It is clear that the integers have at least one property different from those of the natural numbers. If you take any subset of the natural numbers, it has a least member. But there are subsets of the integers that do not—for example, all the even integers.

$$\ldots, -2n, \ldots, -4, -2, 0, 2, 4, \ldots, 2n, \ldots$$

This set has neither a least nor a greatest member.

If, however, you rearrange the infinite set of integers, you can find a one-to-one correspondence with the natural numbers. Start out 0, 1, -1, 2, -2, 3, -3, and so forth, which gives a set with every subset having a *first* member (if not a least member). Then you can "count" it; that is, you can show that the set of integers has the same (infinite) cardinality as the set of natural numbers.

Cantor also looked at the set of rational numbers. (For simplicity in presentation, only the positive rationals will be considered here; the reasoning is the same for the set that also includes the negative rationals). The rationals also have a property that is not shared with either the natural numbers or the integers. Between every two different rational numbers, you can always find as many other rational numbers as you like. For example, between 1/2 and 1/3 you can find their average $(1/2 + 1/3) \div 2$, or 5/12. Between 5/12 and 1/3, you can find their average, 9/24, and so forth, as many times as you like. Between 2 and 3, however, there is no other natural number. So it would appear that there might be more rational numbers than there are natural numbers. That is, the set of rational numbers might not be countable.

Cantor, however, figured out a way to rearrange the rational numbers so that they could be counted, or matched one-to-one with the natural numbers. First he needed to display a set that would include all of the rational numbers. He did this by putting the rational numbers with 1 as a numerator in the first row, the numbers with 2 as a numerator in the second row, and so forth. The display looks like this:

$$\frac{1}{1} \quad \frac{1}{2} \quad \frac{1}{3} \quad \cdots \quad \frac{1}{n} \quad \cdots$$

$$\frac{2}{1} \quad \frac{2}{2} \quad \frac{2}{3} \quad \cdots \quad \frac{2}{n} \quad \cdots$$

$$\frac{3}{1} \quad \frac{3}{2} \quad \frac{3}{3} \quad \cdots \quad \frac{3}{n} \quad \cdots$$

$$\vdots \qquad \vdots \qquad \vdots \qquad\qquad \vdots$$

$$\frac{n}{1} \quad \frac{n}{2} \quad \frac{n}{3} \quad \cdots \quad \frac{n}{n} \quad \cdots$$

$$\vdots \qquad \vdots \qquad \vdots \qquad\qquad \vdots$$

It should be clear that this display includes all of the (positive) rational numbers. It is true that it includes more than one representation of the numbers (e.g., $1/1 = 2/2$). On the other hand, if the natural numbers can be matched one-to-one with this displayed set of numbers, then, by the definition of an infinite set, there will be a subset (in this case, just the rational numbers) that has the same cardinality. If this worries you, however, you can simply skip any rational number that you have met before. All the "repeats" will lie along diagonals that are easy to cross out. To give you a better picture of how the correspondence works, more of the display is shown, in which the "repeats" are removed by shading and natural numbers are circled.

Thus, the one-to-one correspondence for the first 17 natural numbers runs like this.

While this pattern is a bit hard to describe, it is clear that you could extend it indefinitely to show that the natural numbers are indeed in one-to-one correspondence with the rationals.

REASONING WITH BIAS

At this point, it begins to look as if all infinite sets could be put into one-to-one correspondence with each other. While that would not be an unreasonable supposition, it was not to be. For when Cantor turned to the real numbers, instead of discovering a proof that they are capable of being put into one-to-one correspondence with the natural numbers, he discovered the opposite—a proof that they cannot be matched one-to-one.

Furthermore, the unusual method of proof that Cantor used turned out to be the essence of a method that was later used to uncover the most startling facts about mathematics. Therefore, it will get a full treatment in this context and be referred to several times in what follows. It is known as *Cantor's diagonal method*. Basically, it is a form of indirect proof. It is also a nonexistence proof. In short, it does not sit well with the intuitionists.

The real numbers include all of the rational numbers and all of the irrational numbers. While $\sqrt{2}$ has been the only irrational number discussed in this book, there has been occasion to mention π, which also is irrational. The commonest way to think of the real numbers is to picture them as all the points on a line. You know that there are points on a line that are not rational (the point at $\sqrt{2}$ is on the line, for instance).

Another way to picture the real numbers is to think of them as numbers that can be represented by decimals, including decimals that have a countable infinity of digits. It can be shown that such decimals can be put into one-to-one correspondence with the points on a line provided one is careful to eliminate one set of decimals from the set of *all* decimals.

It should be clear that in this discussion decimals are to be treated as numerals—marks on paper or, at least, representations in the mind—rather than as numbers. In ordinary uses of decimals, people think of 0.5 as the common representation for the real number 1/2 (think of 1/2 now as the number, 0.5 as the numeral), $0.33\overline{3}$ or $0.333\cdots$ as

the common representation for the real number 1/3, and 1.4142 · · ·
as the common representation for the real number $\sqrt{2}$. The decimal
0.5 is said to *terminate* because every digit after 5 is 0, which people
do not bother to indicate. The decimal $0.33\overline{3}$ is repeating and nonterminating. (The bar over the last 3 means to keep repeating it to
infinity.) The decimal 1.4142 · · · is also nonterminating, but it is
nonrepeating. It is easy to show that any repeating decimal can be
expressed as the ratio of two natural numbers. Since you know that
$\sqrt{2}$ is irrational, it cannot be a repeater.

The hitch comes as follows. A decimal representation is simply a
way of writing a series. Thus, $0.33\overline{3}$ means the infinite series

$$0 + \frac{3}{10} + \frac{3}{100} + \frac{3}{1000} + \cdots + \frac{3}{10^{n-1}} + \cdots$$

All such series have sums. In particular, the infinite series

$$0 + \frac{4}{10} + \frac{9}{100} + \frac{9}{1000} + \cdots + \frac{9}{10^{n-1}} + \cdots$$

has a sum, and the sum is 1/2. In other words, $0.49\overline{9}$ is equal to 0.5.
This result is often viewed by students learning it for the first time as
a paradox because they are mentally committed to the notion that a
number can be represented in one and only one way by a decimal.
Alas, it is not the case.

The same is true for all other terminating decimals. For example,
$1 = 0.99\overline{9}$, $0.7 = 0.69\overline{9}$, and $0.25 = 0.24\overline{9}$.

While in normal usage it is convenient to represent real numbers
with terminating decimals where possible, Cantor's diagonal method
makes it more convenient always to represent real numbers with decimals that do not terminate (remembering that "terminate" merely
means that 0 repeats infinitely). In this way, there is one and only
one allowable representation of each decimal.

To simplify the presentation of Cantor's proof, apply it to the real
numbers from 0 (but not including 0) to $0.99\overline{9}$, which is the number
usually represented as 1. The number $0.99\overline{9}$ will be included in the
set.

Now in the ordinary relationship of size, this set does not have a first member. That is, the statement x is the first number greater than 0 is meaningless. In 1904, however, in a controversial proof that will be discussed later, Ernst Zermelo proved that all sets of real numbers can be arranged in some way so that there is a first member. In this case, it is easy to see how to do it. Start at the top. Then $0.99\overline{9}$ is the first member.

Recall that Cantor's diagonal method is a form of indirect proof. So you *assume* that the real numbers from 0 to $0.99\overline{9}$ can be matched one-to-one with the natural numbers. Then, you look for a contradiction.

Perhaps $0.99\overline{9}$ is the number matched with 1 in the correspondence, and perhaps it is some other number. In the one-to-one correspondence between the rationals and the natural numbers the correspondence did not match pairs in an expected way. But from Zermelo's theorem, you know that there is at least one way (maybe others) to establish a first match. From the infinite set that remains, you can establish a second match. From the infinite set that remains, you can establish a third match. Therefore, you can set up a sequence of real numbers, which, since you do not know what they are, you can call

$$r_1, r_2, r_3, \cdots$$

For the correspondence to be one-to-one, this sequence must include *all* of the real numbers between 0 and $0.99\overline{9}$.

The diagonal method is a way of finding a real number that is not in the sequence so set up. It is easy to illustrate with a finite example. Consider a sequence (randomly chosen) of 10 decimals, each of which has 10 digits after the decimal point. The diagonal method can be used to find another decimal of the same type that is not in the sequence. Here is such a sequence.

$$0.9278374191$$
$$0.2668251036$$
$$0.0274618363$$
$$0.0374837562$$
$$0.2934635183$$

0.3746567389

0.4835091637

0.1384243845

0.4736784944

0.5537483768

Now proceed to create a new decimal by changing the first digit in the first given decimal, the second digit in the second given decimal, and so forth. For convenience, subtract 1 from the digit to be changed unless the digit is 0, in which case change it to 9. The new decimal is

0.8563550737

Clearly, this decimal was not in the original list, because it is different from the first one in the first place, from the second one in the second place, and so on through the tenth place.

Now return to the infinite sequence of real numbers that starts off r_1, r_2, and r_3. Each of these decimal representations has a countable infinity of places. If there are a countable infinity of decimals as well, you have the same situation as in the 10×10 array, although this is a countable infinity by countable infinity array. You proceed in the same way, changing the first digit in the first, the second digit in the second, and the nth digit in the nth. You can use the same rule for making the change: subtract 1 from every digit but 0 and change the 0s to 9s. The result will be a new decimal that is not in the list.

But the list was supposed to contain *all* decimals from 0 to $0.99\overline{9}$. Even the existence of one that is not in the list is a contradiction. Therefore, the real numbers cannot be put into a one-to-one correspondence with the natural numbers. There are two kinds of infinity after all!

If you have been observant, you may have an objection to this procedure. Suppose that the only possible array that preserves the ordering property is one such that after a certain point the nth digit in the nth decimal is always 1. Then the resulting new decimal will be of the kind that was ruled out to begin with. That is, it will end with an infinity of 0s. It is not on the list, but it is supposed to be not on the list.

There are a number of ways to dispose of this objection. The easiest is to say: "Okay. In that case, here is *another* decimal that is not on the list. This time add 1 to each digit except for 9, and subtract 1 from 9. Or, here is yet another one. Subtract 2 from each digit, changing 1 to 9 and 0 to 8. Or, simply change the digits at random. In any case, you will get a new decimal, one that is not on the original list." And this true. And, if the original rule produced a terminating decimal, none of the others would.

The obvious next question for Cantor was: If there are two infinities, are there more?

It is helpful to have names for things. The cardinality of the set of natural numbers is a countable infinity. Cantor called the cardinality of the set of real numbers the *power of the continuum*. (The *continuum* is another name for the points on a line.) Since the power of the continuum is the (infinite) number of points in a line, perhaps another infinity could be found by looking at the number of points in a plane. But, perhaps not, since the distinguishing characteristic of an infinite set is that it can be matched one-to-one with some of its subsets.

Again, to simplify the presentation, consider just a part of a plane. The argument generalizes in an obvious fashion to all of the plane, but it is harder to write down. The part used in the proof is a square with two sides missing for which, when the plane is assigned coordinates, for each coordinate (x, y) both x and y are between 0 and $0.99\overline{9}$—still excluding terminating decimals.

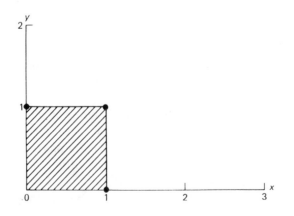

Now consider any point (x, y) in this region. For illustration, consider $x = 0.2937847\cdots$ and $y = 0.7836655\cdots$. To establish a one-to-one correspondence between any point in the region and any point on the line between 0 and $0.99\overline{9}$, you need a method of obtaining a single decimal from a pair of them. Also, the method should be such that you can use it in reverse to obtain the original pair.

Here is one method. To make a single decimal from $0.2937847\cdots$ and $0.7836655\cdots$, choose as your first digit the first digit of the first decimal (x); choose as your second digit the first digit of the second decimal (y); choose as your third digit the second digit of the first decimal; choose as your fourth digit the second digit of the second decimal; and so forth. The result (for the examples given) is

$$0.27983376864575\cdots.$$

This fulfills the first requirement, for the new decimal is one of the ones between 0 and $0.99\overline{9}$. It also fulfills the second requirement, for you can look at any decimal and recover a pair (x, y) from it. For example, the decimal $0.92725892\cdots$ matches with $(0.9759\cdots, 0.2282\cdots)$.

Using this rule of correspondence, each point on the line matches in one and only one way with a point in the region. The rule also matches each point in the region in one and only one way with a point on the line. This is sufficient to establish that the two sets of points have the same cardinality.

What is more, by considering coordinates in three dimensions, (x, y, z), the same method can be used to show that the number of points on a line segment is the same as the number of points in all (mathematical) space; that is, the infinities have the same cardinality.

SETS OF SETS

You still have seen two kinds of infinity only. Cantor, however, was able to find additional infinities with different cardinalities from either the countable infinity or the power of the continuum.

To explore this proof of Cantor's, it is necessary to be somewhat clearer about *sets* and *subsets*. First of all, so that you know exactly what is and what is not a set, a *set* is something about which you know (for, presumably, everything in the universe) whether or not a given object (real or mental) is a member of the set or not. For example, the set of natural numbers is a set because you can tell that 2 is a natural number, $\sqrt{2}$ is not a natural number, 1,092,728,036 is a natural number, the letter A is not a natural number, a button on your shirt is not a natural number, and so forth. The "set" of beautiful women is not acceptable, because you cannot objectively tell which women are beautiful and which are not (even though you may have some internal criterion).

A *subset* of a given set is a set such that none of its members is not also a member of the given set. Thus, the set of numbers less than 10 is a subset of the set of real numbers (but not a subset of the set of natural numbers, for $\sqrt{2}$, say, is less than 10 but not a natural number). It is fairly easy to see that the given set is always a subset of itself. With a little more thought, you can also see that the set with no members at all, called the *empty set*, is a subset of every set.

One way to indicate a set is to list its members in {braces}. Thus, the set of natural numbers less than 10 can also be shown as $\{1, 2, 3, 4, 5, 6, 7, 8, 9\}$. For a small set, such as $\{a, b, c\}$ it is easy to list all of the subsets. These are

$$
\begin{array}{ll}
\{\ \} & \{a, b\} \\
\{a\} & \{a, c\} \\
\{b\} & \{b, c\} \\
\{c\} & \{a, b, c\}
\end{array}
$$

The order that the members are listed in does not make any difference.

Notice that $\{a, b, c\}$ has three members and the set of all of the subsets of $\{a, b, c\}$ has eight members. This is because for each subset you have two choices about each member—whether it is or is not be a member of the subset. So for each member, you choose once to keep it and once to eliminate it when you are obtaining all of the subsets. This means that the number of subsets will be $2 \times 2 \times 2$, or 2^3, or 8 (in this case). Furthermore, for any finite set with n members, the number of subsets of that set will be 2^n. Finally, since 2^n is greater than n for all natural numbers n, there will always be more subsets of a finite set than there are members.

It may be clearer to prove that a set with n members has 2^n subsets using mathematical induction.

Step 1. Show that a set with one member has two subsets. Since every set has the empty set and itself as subsets, a set with one member has these two subsets. And, in fact, there are no other subsets to be found. So the statement is true for $n = 1$.

Step 2. Now assume that for some specific number k, a set with k members has 2^k subsets.

Step 3. Consider a set with $k + 1$ members. The first k members form, by the assumption in Step 2, 2^k subsets of the set. Now take each of these subsets and form a new subset that also includes the last member of the set with $k + 1$ members. You can do this 2^k times. In this way, you produce all of the subsets of the set with $k + 1$ members. Since there were 2^k that did not include the last member and 2^k that did include the last member, the total number will be $2^k + 2^k$ subsets. But this is the same as 2×2^k subsets, which is equal to 2^{k+1} subsets.

Therefore, you have shown by mathematical induction that a set with n members has 2^n subsets. Therefore, any set has more subsets than it has members. Mathematical induction, however, only works for finite sets in this proof. Cantor used his diagonal method to show that this relationship is also true for infinite sets.

Start with a set, which you can call R. There exists at least one set of subsets of R that has the same cardinality as R. (For example, there is the set of each member of R considered as a set. For $\{a, b, c\}$, that would be $\{\{a\}, \{b\}, \{c\}\}$.) Let S be the set of all subsets of R. Since the same rules do not always hold for infinite sets as for finite sets, assume that S also has the same cardinality as R.

Cantor used the diagonal method to show that there must also exist a subset of R that is not is S. Since this is an indirect proof, you are going to show that the assumption that S is in one-to-one correspondence with R contradicts the definition of S as "all subsets of R." But you begin by assuming that both the assumption and the definition are true. Then you can create a one-to-one correspondence between each member of R and each subset in S. This is similar to the attempt to set up the one-to-one correspondence between the natural numbers and the real numbers.

Now create a subset of R which is defined as follows. In the one-to-one correspondence between members of R and members of S, it is possible that the member of R that corresponds to a particular member of S is *also* a member of S. After all, since S is a set of subsets of R, all of the members of sets in S are also members of R. When a member of R is not also a member of the subset to which it corresponds, assign it to another subset, which you can call T.

Since all the members of T are members of R, T is a subset of R. Therefore, according to the assumption that S is all of the subsets of R, somewhere in S the set T must match some member of R in the one-to-one correspondence you have assumed. Is this member—call it r—a member of T or not?

By the definition of T, if r is a member of T it should not be assigned to T. But if r is not a member of T, it should be assigned to T. Since this is a contradiction, T cannot be a member of S. This is another way to say that the set of subsets of a set always has more members than the original set.

(Notice how this proof resembles Grelling's paradox of the heterological adjective. The question "Is r a member of T?" produces the same kind of self-contradiction as the question "Is 'heterological' heterological?". In Cantor's proof, however, the assumption that T is in S is easily denied. In Grelling's paradox, it is difficult to deny any of the assumptions.)

Since for a finite set with n members, the number of its subsets could be denoted 2^n, Cantor denoted the cardinality of the number of subsets of an infinite set in the same way. If, for example, the cardinality of the set of real numbers is c, then 2^c is the cardinality of the number of its subsets. Similarly, if the cardinality of the set of natural numbers is a—another symbol is usually used, the letter *aleph* of the Hebrew alphabet with 0 as a subscript—then 2^a is the cardinality of the set of all subsets of the natural numbers. In that way, the rule that the number of subsets of any set is 2^n, when there are n members in the set, was extended so n could be infinite.

By taking subsets of subsets, you can get to still greater cardinalities. Looking at the finite case again, while there are $2^3 = 8$ subsets of $\{a, b, c\}$, there are $2^8 = 256$ subsets of the set

$$\Big\{\{\ \}, \{a\}, \{b\}, \{c\}, \{a,b\}, \{a,c\}, \{b,c\}, \{a,b,c\}\Big\},$$

which is, of course, the set of subsets of $\{a, b, c\}$.

So the set of subsets of the set of subsets of the natural numbers (call it S_2) has a greater cardinality, 2^{2^a}, than the set of subsets of the natural numbers (S_1). And the set of subsets of the set of subsets of the set of subsets of the natural numbers (S_3) is still larger. And this can continue for a countable infinity of times. These sets are known as the *transfinite cardinals*. Alternatively, the numbers assigned to the sets may be called the transfinite cardinals.

Cantor went on to show that 2^a is equal to c, but the proof is quite complicated and not needed for what follows. He also developed a theory of infinite ordinal numbers, which it is worthwhile to describe briefly.

ORDER IN INFINITY

Ordinal numbers are based on the concept of *order types*. If two sets have the same cardinal number, they can be matched one-to-one. If the matching can be done in such a way that the order of each set stays the same, then the sets have the same order type. In dealing with order types, unlike the earlier work with sets, the order in which the members are listed is important. All finite sets that have the same cardinal number also have the same order type. For example, let the order be the usual order of the natural numbers. Then the sets $\{1, 2, 3\}$ and $\{195, 137, 206\}$ can be matched

Since each finite set of a given cardinality is of the same order type, the total number of order types of the finite sets is the same as the cardinal number a, the cardinal number of the set of natural numbers. In this context, however, it will be called a_0.

(You know that $a = a_0$ because there is one order type for all the sets with one member, one order type for all the sets with two members, and so forth. So there is an order type for each of the natural numbers.)

Now consider the different orders in which a countable infinity of members could be arranged. For convenience, take the countable infinity of natural numbers.

One way is the ordinary way.

$$1, \ 2, \ 3, \ 4, \ 5, \ \ldots$$

You could also interchange the first two numbers.

$$2, \ 1, \ 3, \ 4, \ 5, \ \ldots$$

However, it is possible to set up a correspondence that preserves the order in that case.

Similarly, there is an order-preserving correspondence if you interchange each two numbers.

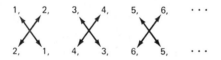

But suppose that you take 1 and put it last in the ordered set.

$$2, \ 3, \ 4, \ 5, \ \ldots, \ 1$$

In that case, the order types are fundamentally different. Since this set contains a last member, there is no way that you can match it with the usual ordering, which does not contain a last member.

Furthermore, if you arrange the set with all the odd numbers before the even numbers

$$1, \ 3, \ 5, \ \ldots, \ 2, \ 4, \ 6, \ \ldots$$

you obtain a set that has still another ordering. This set has no last member and it has a member—2—that has no specific member preceding it.

Cantor showed that the number of different ways to order a countably infinite set was infinite and that this infinite number was different from (and greater than) the number a_0. This number can be called a_1.

The theory of ordinal infinities continues to develop higher and higher infinities, first a_2, a_3, and so forth, and then higher infinities still. Each of these is built on the set of ordinals that precedes it.

Except for a and a_0 it was not clear how these infinite numbers were related to the infinite numbers that start out 2^a, 2^{2^a}, and so forth. For example, is a_1 the same number as 2^a? Cantor thought that it was, but he could not prove it. Since 2^a (or c) is the power of the continuum, this speculation of Cantor's came to be known as the *continuum hypothesis*. In words, it can be stated as: The power of the continuum, or the cardinal number of the set of real numbers, is the same as the number of ways that the set of natural numbers can be ordered. It was not at all clear in Cantor's time whether or not he was right in his speculation, but the problem was finally resolved in 1963. The result was a surprise, which will keep nicely until the next chapter.

THINGS GO WRONG

Cantor's theory of ordinal numbers soon was shown (by Cantor himself) to contain a paradox. He apparently realized this in 1895, but he did not publish it. It was published in 1897 by Cesare Burali-Forti (1861–1931), so it is known today as the *Burali-Forti paradox*. (This sort of thing happens all the time in mathematics. Most of the familiar names of mathematical theorems are mistaken attributions to later developers of the theorems.) The Burali-Forti paradox is sufficiently unpleasant that when it was shown to appear in a system of logic devised by W. V. O. Quine, he felt forced to make major changes in the system.

The Burali-Forti paradox arises from the fact that you can build a new ordinal number from each preceding one. This series of ordinal numbers also has an ordinal number. That ordinal number ought to be somewhere in the series. But like the set T in the proof that the number of subsets is greater than the number of members of a set, this ordinal can be shown not to be in the series of ordinal numbers. This is a contradiction, but in this case, unlike the proof, there is no assumption to contradict—unless the whole theory is abandoned.

Cantor also discovered another paradox, one that is much easier to understand than the Burali-Forti paradox. Although he did not publish this problem either, it came to be known as the *Cantor paradox*.

Cantor had shown that for any set whatsoever, the set of subsets of the set contains more members than the set itself. What about the set of all sets?

Since the set of all sets includes all possible sets, each of its subsets must be members of it. So there cannot be more subsets than there are members of the set of all sets.

Cantor apparently discovered both the Burali-Forti paradox and the Cantor paradox around 1895, but word of them slowly reached the mathematical public. Both paradoxes came at a very awkward time in the history of mathematics. Several mathematicians and logicians had just embarked on programs to show that mathematics was free from contradictions. In all cases, they were using sets freely in their efforts. If set theory contained contradictions (which is what the paradoxes amounted to), then their efforts would be much more difficult than they had expected.

In fact, the whole nineteenth century had seemed to be heading toward a complete and rigorous development of all of mathematics. The discovery of non-Euclidean geometries early in the century had led mathematicians to examine their fundamental assumptions more closely. Augustin Cauchy and Karl Weierstrass had reformed the calculus. Around the middle of the century, in England, George Boole and Augustus De Morgan had begun the development of symbolic logic, which led to a much more careful examination of the basis of the logic that was applied in mathematics. In the 1890s, Giuseppe Peano in Italy was developing arithmetic on an axiomatic basis, and Gottlob Frege in Germany was working to develop mathematics from pure logic.

In 1900, David Hilbert (1862–1943), in a famous speech before an international conference of mathematicians, summed up what he hoped could be accomplished in the twentieth century by identifying 23 outstanding unsolved problems of mathematics. Problem Number Two was to show that mathematics is consistent. He was aware of paradoxes in set theory (Cantor had mentioned the Burali-Forti paradox to him in 1895), but he had no idea how difficult Problem Number Two was going to be.

(Problem Number One, by the way, had two parts. The first part was to determine the truth or falsity of Cantor's continuum hypothesis. The second was to show, as was assumed in the proof that the real numbers are not countable, that there is a way to find a first number for every set of real numbers.)

In the very next year after Hilbert's speech, Bertrand Russell (1872–1970) added another paradox to the list, posing still graver problems for set theory. There is a well-known story among mathematicians about how Russell passed along the bad news.

Gottlob Frege had been working on developing arithmetic from logic for a quarter of a century. He planned a two-volume presentation of his whole development and had the first volume published and the second completed when he got the letter from Russell. Readers of the second volume, consequently, found a footnote at the end of the volume that began

A scientist can hardly meet with anything more undesirable than to have the foundation give way just as the work is finished. In this position I was put by a letter from Mr. Bertrand Russell as the work was nearly through the press.

Frege had written earlier that mathematicians must "face the possibility that we may still encounter a contradiction that brings the whole edifice down in ruins. For this reason I have felt bound to go back rather further into the general logical foundations of science . . ." It is somewhat ironic that Frege was a strong advocate of examining the foundations of mathematics, for this is where *Russell's paradox* occurs.

Russell had heard rumors of Cantor's paradox. In thinking about it, he produced his own. Russell's paradox can be shown to be related to the proof that the set of subsets of a set contains more members than the set itself. It should also remind you of Grelling's paradox, but Grelling devised his paradox later and was probably influenced by Russell's paradox.

Russell's paradox is particularly striking because it deals with just the basic concept of sets: membership. Recall that a set is defined as something about which you know for any given object whether or not the object is a member of the set. In this definition, *member* is an undefined term. Intuitively, however, we think we understand what it means.

For one thing, normally, a set is not a member of itself. For example, consider the set of rivers of the United States. It contains the Mississippi, the Missouri, the Columbia, the Hudson, and many others as members. It does not contain the set of rivers of the United States as a member. Call a set that does not contain itself as a member *normal*. The set of natural numbers is another normal set. In fact, it appears that most sets are normal.

But there do exist *abnormal* sets, sets that contain themselves as members. A mathematician might, for example, refer to the set of sets with more than one member. Since this set contains both the set of rivers in the United States and the set of natural numbers, it also contains the set of sets with more than one member. In other words, it is an abnormal set.

Now use this property to divide all sets into two categories: normal and abnormal. Because you can always tell whether or not any particular set is a member of a given set (from the definition of *set*), the two categories are mutually exclusive and every set can be put into one category or the other.

Now consider the set of all normal sets. Is this set normal or abnormal?

If it is abnormal, then it has itself as one of its members. But since its members are normal, this is a contradiction.

If it is normal, then it does not have itself as a member. But, since this is the set of *all* normal sets, it must have itself as a member. This is also a contradiction.

Now this means that the set of all normal sets does not meet the criterion established in the definition of *set*. A set is a collection of objects (real or mental) such that you can always tell, when presented with any object, whether or not the object is a member of the collection. But when you are presented with the set of all normal sets, you cannot tell whether or not it is a member of the set of all normal sets.

Notice the parallelism with Grelling's paradox.

Grelling	Russell
Adjectives either describe themselves or do not describe themselves.	Sets either include themselves or do not include themselves.
Call adjectives that do not describe themselves heterological.	Call sets that do not include themselves normal.
Is *heterological* heterological or not?	Is the set of normal sets normal or not?

While Grelling's paradox is an exact parallel to Russell's paradox, Grelling's paradox is about *words*. Russell's paradox results from the kind of mathematical statement that mathematicians make every day. For example, a mathematician may think "all the numbers that are not equal to a" and write this as $\{x : x \neq a\}$, where the colon means *such that* and $\neq$ means *is not equal to*. Similarly, he or she may think "all the sets that are not members of A," which is written $\{X : X \nsubseteq A\}$, where $\nsubseteq$ means *is not a member of*. For numbers, you still have a well-defined expression when you substitute x for a in the first expression. The set of numbers described by $\{x : x \neq x\}$ is the empty set, since x is always equal to itself. But when you substitute X for A in $\{X : X \nsubseteq A\}$, you get Russell's paradox, for $\{X : X \nsubseteq X\}$ is "the set of all sets that are not members of themselves." Yet $\{X : X \nsubseteq X\}$ is the kind of expression that could crop up in mathematics at any time. If mathematics is to be consistent, then there needs to be some rule that prevents one from encountering such paradoxes.

REPAIRING THE DAMAGE

In fact, it was to avoid this sort of thing that Poincaré introduced the concept of impredicative definitions. The definition of the set of all normal sets is impredicative. Using Poincaré's rule, it would not be allowed into mathematics. Unfortunately, there is a lot of very basic mathematics that also relies on impredicative definitions. When Hermann Weyl made a serious effort to redevelop mathematics without using impredicative definitions, he failed badly. For example, the concept of the *least upper bound*, which is essential to calculus, is an impredicative definition. The least upper bound is the smallest number that is larger than all of the members of a particular set. For example, 10 is the least upper bound of the set of numbers less than 10.

Even in this book, which does not go very far into the theoretical foundation of mathematics, there is a need for impredicative definitions. For example, in Chapter 2, the definitions of *sequence* and the *maximum function* MAX(x, y) are both impredicative. In showing that $1 - 1/2 + 1/3 - 1/4 + \cdots$ has a sum, the procedure of taking the power of $1/2$ that is the greatest number less than each term is an analog of finding a least upper bound.

Also, Cantor's proof that the number of subsets of an infinite set exceeds the number of members of the set would fall apart, since it is based on an impredicative definition.

Ruling out impredicative definitions probably would eliminate the contradictions from mathematics, but the cost is too great.

Bertrand Russell suggested another problem with *impredicative*. By definition, a property is impredicative if it applies to itself. Therefore, it is predicative if it does not apply to itself. Is the property of being impredicative itself impredicative or not? (This is another analog of Grelling's paradox.)

Russell also proposed another way to eliminate the paradoxes from mathematics. In his system, called the *theory of types*, a hierarchy is set up. For sets, members are on the first rung of the ladder, sets of ordinary members are on the next rung, sets of sets on the third rung, and so forth. Similarly, you could call a particular pencil *type 0*, a statement about that pencil *type 1*, a statement about statements about pencils *type 2*, and so forth. For example, the statement "Any statement about the pencil on my desk is false" is *type 2*. (And the preceding sentence is *type 3*.)

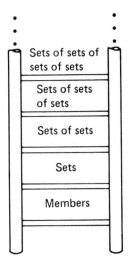

In the theory of types, a statement or a set of *type k*, where *k* is some particular natural number, cannot be applied to another statement or set of *type k*. This produces unending complications.

For example, since the rational numbers are defined in terms of ratios of natural numbers, rational numbers and natural numbers are of different types. As the theory has been stated so far, that would mean that you would need separate proofs for the natural numbers and the rational numbers—which, sometimes, you do, but not *always* in ordinary mathematics.

To get around this and other problems, Russell introduced what he called *axioms of reducibility*. These were designed specifically to permit mathematics to be developed from a logic that included the theory of types. Since they had no other purpose or use, and furthermore seemed to go against intuition, Russell eventually abandoned the idea.

In 1926, F. P. Ramsey invented the *simple theory of types* to solve the paradoxes, but it was so nonintuitive that it has been called "theological" by mathematicians.

Thus, these attempts to solve the paradoxes all turned out to involve either paradoxical notions themselves or to be so artificial that most mathematicians rejected them.

Another solution to the paradoxes was to adopt intuitionism. In an intuitionistic point of view, the set of all normal sets poses no problem, since you cannot construct it. Therefore, it does not exist. The number of the subsets of the set of all sets is eliminated on two grounds; not only can it not be constructed, but the proof that it will contain more subsets than members is also unacceptable. Unfortunately, intuitionism is also apparently incapable of reconstructing much useful mathematics (although, in theory, it is not impossible for intuitionism to rebuild most really useful mathematics).

A FORMAL PROGRAM

In the 1920s David Hilbert attacked Problem Number Two from the list of 23 he presented in 1900. Dissatisfied with the attempts of others to prove the consistency of mathematics, he developed a program for reconstructing mathematics in a way that could lead to results that he would find acceptable.

He began by disassociating mathematics from reality. Since the reasoning process used in mathematics does not depend on what you are reasoning about, leave the subject out of the picture entirely. Consider that you are just dealing with rules for manipulating symbols.

Hilbert believed if you set up the rules carefully, you could both avoid paradoxes and still develop all that was worthwhile in mathematics. It was not essential that the rules be intuitively obvious; they just had to work. Furthermore, he believed that, by reasoning about the system from *outside* the system, he—or another mathematician— could prove that the system was consistent. That is, you would never end up with 2 = 1 as a result of a valid proof. He also believed that the system could be shown to be complete. A complete system would be capable of proving anything that was, in the sense of the system, true.

Hilbert's program, known as *formalism*, scored some early successes, which encouraged him greatly. Some parts of mathematics were shown to be both consistent and complete. But, for example, it could not be shown that arithmetic was complete and consistent if it included multiplication.

Hilbert divided mathematics into two levels and then added a third. Level One was ordinary mathematics, the casual way it had developed since Pythagoras and Euclid. Level Two was a system of formal symbols, very strictly defined, that could be manipulated by strictly defined rules. Level Two did not describe anything, but it was inspired by Level One. (In other words, a mathematician looking closely at the system in Level Two would recognize that it was a formalization of mathematics.) Level Three would be an informal theory about Level Two. In Level Three, reasoning would be conducted with almost the freedom that had been used in developing Level One. However, in Level Three, completed infinities were outlawed (although allowed in Level One). Also, no existence proofs could be used, although some types of nonexistence proofs were acceptable. Level Three is called *metamathematics*.

It should be clear that had Hilbert's formalist program succeeded (it did not), the problem of the paradoxes would be resolved. On the other hand, adoption of Hilbert's system by itself did nothing to eliminate the paradoxes directly, although a suitable development of Level Two could eliminate the known paradoxes from that level. For example, one of the symbolic statements in Level Two could mean something like "For no X does $\{X : X \nsubseteq X\}$ exist." (Loosely, since the symbols mean nothing and existence is not assumed.)

Working mathematicians, if they paid any attention to these concerns, found little to help them in any of these approaches, but they did take to Cantor's set theory. Many used Cantor's notions informally, as Cantor himself had. This use of sets came to be known as *naive set theory*. It was naive, because it led to paradoxes, but it was a useful language and the nonparadoxical results were helpful.

More careful mathematicians, however, felt the need for a version of set theory that was both powerful enough for mathematical use and restrained enough to avoid the known paradoxes. To provide this, various mathematicians put forward axiomatic developments of set theory. The axiomatic treatments were developments of set theory modeled on the method used by Euclid but informed by critiques of Euclid that had occurred in the nineteenth and twentieth centuries. The first of such systems was developed in 1908 by Ernst Zermelo. Later mathematicians, including Abraham Fraenkel, John von Neumann, Paul Bernays, and Kurt Gödel, fiddled with the axioms in various ways. While various formulations are slightly different, the main ones that are used by mathematicians have been shown to be equivalent.

Morris Kline, in his book *Mathematics: The Loss of Certainty*, gives a formulation of this kind of axiom system entirely (almost) in English. Since the formulations of mathematicians are almost always entirely in symbols, it is helpful to look at Kline's version. Kline does not intend this to be complete, and, of course, a formulation in symbols is needed for use in mathematics.

1. Two sets are identical if they have the same members. (Intuitively, this defines the notion of a set.)
2. The empty set exists.
3. If x and y are sets, then the unordered pair $\{x, y\}$ is a set.
4. The union of a set of sets is a set.
5. Infinite sets exist. (This axiom permits transfinite cardinals. It is crucial because it goes beyond experience.)
6. Any property that can be formalized in the language of the theory can be used to define a set.
7. One can form the power set of any set; that is, the collection of all subsets of any given set is a set. (This process can be repeated indefinitely; that is, consider the set of all subsets of any given set as a new set; the power set of this set is a new set.)
8. The axiom of choice.
9. x does not belong to x.

The axiom of choice will be discussed in detail in Chapter 7. Roughly, it says that from an infinite set you can select a subset with a particular property even when you have no particular method to select the members of the subset.

Axiom 9 from Kline's version rules out Russell's paradox. The axioms provide no method of discussing such entities as the set of all sets or the sets of all ordinals, so Cantor's paradox and the Burali-Forti paradox cannot occur. At least, no one has figured out a way for these paradoxes to occur in axiomatic set theory.

There could be, however, other paradoxes that turn up unexpectedly. (In fact, Skolem's paradox—Chapter 7—does occur, but set theorists solve that by claiming that it is not really a paradox.) The axioms of set theory, like Level One of Hilbert's program, are used because they are in accord with intuitive ideas about how sets should work. The rest of mathematics can be built upon this sandy soil, and frequently this has been done for large parts of mathematics. That it does not reject any nonparadoxical parts of mathematics is the best thing axiomatic set theory has going for it.

All in all, how do the paradoxes discussed in this chapter stand today?

None of them has been resolved by thinking the way mathematicians thought until the end of the nineteenth century. To get around them requires some reformulation of mathematics. Most reformulations of mathematics, except for axiomatic set theory, result in the loss of mathematical ideas and results that have proven to be extremely useful. Axiomatic set theory explicitly eliminates the known paradoxes, but cannot be shown to be consistent. Therefore, other paradoxes can occur at any time.

6
The Limits of Thought?

It is safe to say that no attitude has been completely successful in answering the fundamental questions, but rather the difficulties seem to be inherent in the very nature of mathematics.

Paul C. Cohen

By 1930, David Hilbert's program for showing that mathematics, suitably formulated, was free from contradiction despite the paradoxes, was looking good. Some successes in parts of mathematics had been obtained, and the methods being used were improving steadily.

One year later, it was clear that Hilbert's program could not succeed. Disagreements about how to eliminate contradictions were replaced by discussions of how to live with contradictions in mathematics.

In 1931 Kurt Gödel figured out a way to map the paradoxes of language into mathematics. No longer could these paradoxes be dismissed as some peculiar feature of language, as one might say that philosophical discussions of whether or not "Pegasus" stands for a real thing are inconsequential or meaningless.

Gödel was inspired by The Liar and by Richard's paradox. (Recall that Richard's paradox was also the basis for Berry's paradox of the least number that cannot be named in 100 letters or fewer.) In 1905 Jules Richard used Cantor's diagonal method to describe a number that, in one sense of the paradox, cannot be named. The paradox is deliberately set up to resemble Cantor's proof that the real numbers are not countable.

Real numbers can be, as in Berry's paradox, defined by expressions in English (Richard, of course, used French) such as "the square root of two," "the first prime number," and "the ratio of three to one." Each of these real numbers is defined in a finite number of words. Now consider the set of all real numbers that can be defined in a finite number of words. As in Cantor's proof, these real numbers can also be represented uniquely by nonterminating decimals (with the convention that decimals that terminate in the ordinary representation will instead repeat a countably infinite sequence of 9s).

Now arrange the decimal representations in the same order as the set of definitions in English when these definitions have been put in alphabetical order. This set consists of all the real numbers that can be defined in a finite number of words.

Use this arrangement or list to form a new decimal. Its first digit should be altered from the first digit of the first decimal in the list in some way (for example, change it to a 5 if it is not a 5 and change it to a 6 if it is a 5); the second digit should be altered from the second digit of the second decimal in the list; and, in general, the nth digit should be altered from the nth decimal in the list. The result is a decimal representation that does not appear on the list.

The list, however, was to have included all real numbers that could be described in a finite number of words. The procedure just given uses a finite number of words to define a real number that is not on the list.

A MAP OF ITSELF

Richard's paradox can be dismissed as others have been by saying that it relies upon language. In this case, you have the vague notion of a name in English for a real number. Gödel eliminated that vagueness and produced results that could not be dismissed.

The notion of *mapping* is central to mathematics. You are used to the idea of a map as a drawing or model that represents part of the real world. If you are familiar with analytic geometry, you know that pairs of numbers can be represented by points, that is, a mapping from pairs of numbers to points in a plane. This idea can be extended to represent equations in two variables as curves in the plane (or vice versa) or to represent the complex numbers, such as $2 + 3i$, as points in the plane. The important thing about a map is that there must be a correspondence between two sets of objects in such a way that when you are given one set (for example, a road map) you can find the other (for example, a highway).

(There is a minor paradox concerning maps of the ordinary—not mathematical—kind. You can get more details into and measure distances better on a large map than on a small one. From that point of view, the larger a map, the better it is.   But, as Lewis Carroll pointed out, this rule implies that a very good map would be one that is the same size as the country being mapped.)

Gödel's method begins by mapping an axiomatic system into the set of natural numbers.   Since 1931, people have modified Gödel's original system in various ways, but the essential notion does not change.   A formal axiomatic system, such as axiomatic set theory, can be thought of as rules for dealing with symbols.   The number of symbols used are finite (or can be set up to be finite), so each symbol can be assigned a natural number.   These numbers are combined in such a way that each statement in the system will have a unique natural number.   Furthermore, each sequence of statements is also assigned a number by the system, one that is different from any of the numbers of the statements.   Finally, all the assignments are given in such a way that if you are given some natural number, you can recover the symbols for the statement or the sequence of statements that the number codes.

In this way, Gödel mapped the axiomatic system that had been used by Bertrand Russell and Alfred North Whitehead in *Principia Mathematica* into the set of natural numbers just as Richard mapped the real numbers into their names in a natural language.   The *Principia Mathematica* axioms are extensive.   For an illustration of the method, look at a much simpler system.   The principles of *Gödel numbering* (as the mapping is called) apply to any formal system—that is, to any system in which the axioms are taken to be rules that apply to specific symbols.

The axiomatic system to be mapped is a simplification of a system first put forward by Giuseppe Peano to describe what in this book have been called the whole numbers (the natural numbers and zero). Although this version is too simplified to derive rigorously the properties of the whole numbers, it can be used to derive some of them, and it serves to illustrate Gödel numbering.

Each axiom is stated as a collection of typographical symbols. These typographical symbols are supposed to have no intrinsic meaning, but, since the ideas for them come from properties of the whole numbers, it is possible to interpret them as meaning something about whole numbers.   Such liberties are, of course, strictly forbidden to logicians. To avoid setting up a lot of formal logical rules, you are allowed to use only two rules to reason with the axioms.

Here are the axioms.

1. $(0 \in W)$
2. $(a \in W) \rightarrow (Sa \in W)$
3. a. $(Sa = Sa') \rightarrow (a = a')$
   b. $(a = a') \rightarrow (Sa = Sa')$
4. $\sim(Sa = 0)$
5. $(P0 \ \& \ (Pa \rightarrow PSa)) \rightarrow (Pa)$

Here are the interpretations that you may use, which assume such meanings as

(= left parenthesis
0 = zero
$\in$ = is a member of
W = the set of whole numbers
$\rightarrow$ = implies
$\sim$ = not

and so forth.

The following is a verbal description of the axioms just presented:

1. Zero is a whole number.
2. If a number is a whole number, then its successor is also a whole number.
3. a. If two successors are equal, the numbers are equal.
   b. If two numbers are equal, the successors are equal.
4. Zero is not the successor of any number.
5. If zero has a property and if when a number has the property its successor also has the property, every whole number has the property (via mathematical induction).

The reasoning to be used revolves around two familiar rules.

Rule A. A number can be substituted for a variable or a variable for a variable in a statement if the same substitution is made throughout.

Rule B. If you know *this* is "true" and you know "if *this*, then *that*," you also know *that* is "true."

In applying Rule A to the axioms, 0 and S0 are numbers and a and a' are variables.

More formally, these rules are

Rule A. The symbol 0 or any number of repetitions of S followed by 0 can replace a or a followed by any number of repetitions of ' in a statement provided that the same replacement for each occurrence of a or a followed by any number of repetitions of ' is made. Also, a or a followed by any number of repetitions of ' can also be replaced by a followed by any number of repetitions of ' or by a in a statement provided that the same replacement for each occurrence of a followed by any number of repetitions of ' or of a is made.

Rule B. If $P \rightarrow Q$ is an axiom and P is an axiom, you can obtain Q. Furthermore, if you have obtained both X and $X \rightarrow Y$ (where X and Y are statements), you can obtain Y.

A *statement* in this system is anything in parentheses, or in parentheses and preceded by $\sim$; and P or P followed by any number of repetitions of ' indicates a statement that includes a specific number or variable (which is shown by writing the number or variable after P or after the repetitions of ').

A proof in this system consists of a sequence of statements, some of which are axioms and the rest of which are derived from the axioms by using Rule A and Rule B. By limiting proof in this way, the system is quite weak, but remember that the only purpose here is to demonstrate how Gödel numbering works. There is no serious effort to use a formal system to develop new theorems (or proofs of many familiar theorems). Explanatory notes, which are in the column to the right of the proofs, are not part of the proofs themselves.

Here is a proof.

T1: $(a \in W) \to (Sa \in W)$ Axiom 2

    $(0 \in W) \to (S0 \in W)$ Substitution of 0 for a (Rule A)

    $(0 \in W)$ Axiom 1

    $(S0 \in W)$ Rule B

The informal interpretation would be that the successor of 0 is a whole number, which you can think of as 1.

T2: $(a \in W) \to (Sa \in W)$ Axiom 2

    $(S0 \in W) \to (SS0 \in W)$ Rule A

    $(S0 \in W)$ T1

    $(SS0 \in W)$ Rule B

You can think of SS0 as 2.

In a full proof of T2, the third line would be replaced by the proof of T1.

T3: $\sim(Sa = 0)$ Axiom 4

    $\sim(S0 = 0)$ Rule A

In other words, $1 \neq 0$, or 1 does not equal 0.

This is enough for you to see what a proof is. If is just a sequence of formulas that are formed from other formulas according to definite rules. Now you are going to map this system (which describes, in a limited way, the whole numbers) into the natural numbers. For the map, you use the ordinary representation of natural numbers as 1, 2, 3, . . . instead of S0, SS0, SSS0, ... Also, you are allowed to remember everything you know about numbers in the map (in principle, however, you cannot remember anything you know about numbers when dealing with the axiomatic system).

First, assign a number to each of the typographical symbols. For reasons that will become clear, use odd numbers only.

(	1
0	3
$\in$	5

W	7
)	9
a	11
$\rightarrow$	13
S	15
=	17
'	19
~	21
P	23
&	25

Now you could write, say Axiom 4, as

$$21115111739$$

but that would be ambiguous. How would you know whether that meant

$$21 \ 1 \ 15 \ 11 \ 17 \ 3 \ 9$$

(which is Axiom 4) or

$$21 \ 11 \ 5 \ 1 \ 11 \ 7 \ 3 \ 9$$

which is the following mess:

$$\sim a \in (a\text{W}0)$$

You need a system of separating out the symbols in a way that still produces a number. The expression 21 1 15 11 17 3 9 will not do because a space used in that way is not part of a number.

Gödel based his system on a theorem of arithmetic that says that a natural number greater than 1 is either a prime or it can be shown as a product of prime numbers in essentially only one way. For example, 12 is not a prime but can be shown as the product of primes: $12 = 2 \times 2 \times 3$, which is usually written as $2^2 \cdot 3$. While 12 is also equal to 3×4 and 6×2 and 12×1, each of these involves numbers that are not primes.

The Gödel number of a statement is found by taking the primes in order, using the code for each symbol in order as an exponent, and finding the product. For example, the codes (using spaces "illegally") for Axiom 1, or $(0 \in W)$, are 1 3 5 7 9, so the Gödel number for Axiom 1 is

$$2^1 \cdot 3^3 \cdot 5^5 \cdot 7^7 \cdot 11^9$$

and the Gödel number for Axiom 4 is

$$2^{21} \cdot 3^1 \cdot 5^{15} \cdot 7^{11} \cdot 11^{17} \cdot 13^3 \cdot 17^9$$

Now these are big numbers. But they are clearly natural numbers that could be calculated if necessary. (There are other ways of defining Gödel numbers that give somewhat smaller—although still big—numbers. This way is close to Gödel's original formulation.)

In fact, the actual Gödel number for Axiom 1 in this system is found by multiplying

$$2 \cdot 27 \cdot 3125 \cdot 5764801 \cdot 25937424601$$

which goes so far beyond the ability of my hand-held calculator to compute, I'll just stop there.

It should be noted that all Gödel numbers in this system are even, since they all involve a power of 2. The power of 2 will always be odd, but the Gödel number itself will be even. Thus, not every natural number is a Gödel number.

Now a proof is just a sequence of statements. To map the whole formal system into the natural numbers, you want to be able to assign a single Gödel number to each proof; that is, to each sequence. The trick is to use the same trick. Make the Gödel number of the first statement in the sequence the exponent of 2, the Gödel number of the second statement the exponent of 3, the number of the third statement the exponent of 5, and so forth; then multiply everything together.

Thus, for the proof of T3, which consists of only two statements, you would have the Gödel number of Axiom 4:

$$2^{21} \cdot 3^1 \cdot 5^{15} \cdot 7^{11} \cdot 11^{17} \cdot 13^3 \cdot 17^9$$

Gödel number of second (and last) statement in T3:

$$2^{21} \cdot 3^1 \cdot 5^{15} \cdot 7^3 \cdot 11^{17} \cdot 13^3 \cdot 17^9$$

Gödel number of the proof:

$$2^{2^{21} \cdot 3^1 \cdot 5^{15} \cdot 7^{11} \cdot 11^{17} \cdot 13^3 \cdot 17^9} \cdot 3^{2^{21} \cdot 3^1 \cdot 5^{15} \cdot 7^3 \cdot 11^{17} \cdot 13^3 \cdot 17^9}$$

While this is a very big number indeed, it is still a natural number.

Note that the exponent of 2 in the Gödel number of a proof is always even (since it is a Gödel number itself) while the exponent of 2 for a statement is always odd. Thus, you can quickly tell (in theory only) whether or not a particular number stands for a statement or a proof. Not every natural number produces a statement or a proof when translated back into the formal system (all Gödel numbers of proofs are also even), but this is not required in Gödel's proof.

Notice that Gödel's mapping was used in this case to map an axiomatic system that describes the whole numbers back into the very system that is being described. (The natural numbers that are Gödel numbers are also whole numbers.) It is like a snake swallowing its tail.

THE MATHEMATICAL LIAR

Okay, now that you have this formidable system, what can you do with it?

You can use it to show how a version of The Liar can be derived from the axioms of the whole number system. The particular version is a numerical way to say: "yields a false conclusion when appended to its own quotation" yields a false conclusion when appended to its own quotation. This is Quine's paradox. The method of derivation is related to Cantor's diagonal method, which is also used in Richard's paradox.

First of all, notice that the rules for getting a Gödel number can be expressed entirely in numbers. For example, Rule B becomes something like

Rule B. If X is a Gödel number of a statement and $X \cdot p^{13} \cdot Y$ is also a Gödel number of a statement (where p is a prime), then Y is a Gödel number of a statement.

With such rules, you can express the entire axiomatic system as a part of arithmetic. In fact, you can show that if you can compute the Gödel number of a particular proof by following the rules for computing Gödel numbers from the axioms, then that statement can be proved.

What Gödel did, however, was to compute the Gödel number of a statement that says "This statement cannot be proved."

The method of doing this is fairly subtle. In the following description, some of the subtleties are ignored for the sake of making the proof plausible without making it incomprehensible.

If you have the Gödel number x of a proof and translate it back into the original axiomatic system, you get a sequence of statements. Actually, you get first a sequence of Gödel numbers of statements, which you then have to translate one step further to get the statements in the original language. The last number in the sequence of Gödel numbers is the number of the statement proved by the sequence. Call it y. You can then identify a pair of numbers (x, y) in which x is the number of the proof and y is the number of what was proved.

If the rules of the derivation are arithmetical rules that apply to Gödel numbers, then you can determined by arithmetic when you are given a pair of numbers (x, y) whether or not x is the number of the proof of y. This means also that there is some statement $P(x, y)$ in the axiomatic system that describes this relationship. (This is true because the system can be shown to be strong enough to have statements for all systems you can calculate; this topic, however, is beyond the scope of this book.) Understand that the statement $P(x, y)$ can be represented in the system, but that does not mean that if you are given y there always exists a number x such that $P(x, y)$ holds. On the other hand, if you are given x—and it is a legitimate proof—you can find y easily.

Now consider this process. You can substitute a number for a variable in a statement of the axiomatic system (this is Rule A). Following the idea of the Quine paradox, you can substitute the

Gödel number of a statement for a variable of the statement. For example, in proving T3, or $\sim(S0 = 0)$, you have substituted 0 for x in Axiom 4. But you could have substituted the Gödel number of Axiom 4, which in the axiomatic system is expressed by $2^{21} \cdot 3^1 \cdot 5^{15} \cdot 7^{11} \cdot 11^{17} \cdot 13^3 \cdot 17^9$ S's, followed by a single 0. This would be a hard statement to write out in full, but there is nothing theoretically impossible about it.

Since substitution is one of the rules of the axiomatic system, there is a statement in the system that describes substitution of the Gödel number of a statement for a variable of the statement; call the statement that describes this process $Q(z, y)$, where z is a variable. In terms of Gödel numbers, $Q(z, y)$ means that y is the Gödel number of the statement obtained by substituting the Gödel number of the statement whose Gödel number is z in place of any variable in the statement. This sounds complicated, but it can be unraveled if you read through it carefully. Here is an example with specific numbers. Axiom 4 is $\sim(Sa = 0)$. The Gödel number for Axiom 4 is

$$2^{21} \cdot 3^1 \cdot 5^{15} \cdot 7^{11} \cdot 11^{17} \cdot 13^3 \cdot 17^9$$

Abbreviate that number for the moment as G_4, which can be represented by a lot of S's followed by a 0. Then $Q(G_4, y)$ tells you to substitute G_4 for a in $\sim(Sa = 0)$, giving the statement $\sim(SG_4 = 0)$. The Gödel number of this last statement will be y. The number y starts out

$$2^{21} \cdot 3^1 \cdot 5^{15} \cdot 7^{15} \cdot 11^{15} \cdot 13^{15} \cdot 17^{15} \cdot 19^{15} \cdot 23^{15} \ldots$$

and continues in that way with G_4 repetitions of the exponent 15. The statement whose Gödel number is y, which happens to mean that the successor of G_4 is not 0, in this case is true.

The objective now is to develop a statement in the system that says some mathematical form of "this statement is not provable."

Consider the statement

$$\sim P(x, y) \& Q(z, y)$$

The first part of this statement means that it is not true that the sequence represented by the Gödel number x is a proof of the statement represented by the Gödel number y (this is $\sim P(x, y)$). The second part of the statement is that y is the Gödel number of the statement obtained by substituting the Gödel number of some statement z for any variable in statement z (identified by its number). This is $Q(z, y)$. The symbol & means that both of these conditions are to exist at the same time.

Since $\sim P(x, y)$ & $Q(z, y)$ is a statement in the axiomatic system, it has a Gödel number. Call this number g. By the substitution rules, you can substitute g for z to obtain

$$\sim P(x, y) \ \& \ Q(g, y)$$

which means "It is not true at the same time that the sequence represented by x is a proof of y and that y is obtained by substituting g for variables in the statement whose Gödel number is g." Now it should be clear that y exists, since you can always substitute a number for variables (although there are rules that must be followed, which have been spelled out in this discussion). It should also be clear from the definition of $Q(z, y)$ that y is the Gödel number of $\sim P(x, y)$ & $Q(g, y)$. Therefore, x cannot exist, since $\sim P(x, y)$ is one of the conditions in the statement. But x is the proof of y and y is the number of the whole statement

$$\sim P(x, y) \ \& \ Q(g, y)$$

In other words, $\sim P(x, y)$ & $Q(g, y)$ is the mathematical version of The Liar. It is a statement X that says "X is not provable."

Therefore, if X is provable, it is not provable, a contradiction. If on the other hand, X is not provable, then its situation is more complicated. If X says it is not provable and it really is not provable, then X is true, but not provable.

Rather than accept a self-contradictory statement, mathematicians settle for the second choice. That is, there are true statements (e.g., $\sim P(x, y)$ & $Q(g, y)$) in this axiomatic system that cannot be proved.

In fact, this result suggests that any unproved statement, such as Goldbach's conjecture, for example, may turn out to be a statement that cannot be proved. After all, if you know that there are some statements that cannot be proved, there may be others. In fact, this turns out to be the case, as is discussed later in the chapter.

It may occur to you to think that since $\sim P(x, y)$ & $Q(g, y)$ cannot be proved, then perhaps its negation could be proved. But if the statement $\sim(\sim P(x, y)$ & $Q(g, y))$ could be proved, then you could prove a false statement in the axiomatic system. If so, the axiomatic system would be inconsistent. So there exists, for any axiomatic system that is strong enough to derive the natural numbers, a statement such that neither its truth nor its falsity can be proved, unless the system is inconsistent. The last sentence is Gödel's theorem—technically, *Gödel's incompleteness theorem*, since he is responsible for several other important results of modern mathematics.

It immediately follows from Gödel's theorem that Hilbert's program could not succeed, since any axiomatic system strong enough to derive arithmetic would suffer from the same incompleteness as arithmetic. This is because the proof of the theorem relies on only a few properties that most axiomatic systems of any complexity at all share.

DECISIONS, DECISIONS

Gödel had shown a way to tackle axiomatic systems as a whole—although he was preceded in some respects in this kind of approach by Leopold Löwenheim and Thorlaf Skolem (between 1915 and 1930). Others soon began a thorough study of problems that could not be decided within systems. In a way, such problems are analogs of the ancient Greek problems of doubling the cube, squaring a circle, and trisecting an angle. That is, these problems of antiquity were unsolvable given the condition imposed—which was that they be solved with a straightedge (a ruler with no marks on it) and compasses (of the Greek kind, which collapsed when you moved them from one place to another). The Greeks themselves had shown that the problems could be solved without these restrictions. In the nineteenth century, it was shown that the problems could not be solved within the restrictions.

There was something different about the new class of problems that was being investigated, however. Instead of very restrictive methods, such as the straightedge and the compass, mathematicians began to show that problems existed that could not be solved by very general methods. In fact, several mathematicians described such general methods that they came to believe that these methods were equivalent to all possible methods. This belief, known as the *Church-Turing thesis* (after Alonzo Church and Alan Turing), can be more specifically stated as "The only method of calculating a number or set of numbers in a finite number of steps is to use the class of methods that have been identified already." Of course, it is not known whether or not the Church-Turing Thesis is true or not. The evidence for it, however, takes 50 to 60 closely written pages just to state, and the evidence against it is zilch. Consequently, most mathematicians believe that the Church-Turing thesis is true (although it cannot be proved to be true).

In what follows, when it is said that something cannot be solved or cannot be proved, this means that it cannot be solved or proved according to the methods in the class referred to in the Church-Turing thesis. There are several formulations of these methods. The most intuitively appealing is the *Turing machine*, a mental construct invented by Alan Turing that is based on the idea of a machine that computes according to a few simple rules.

A Turing machine is pictured as a machine that is processing an infinitely long strip of paper that has been marked into squares. Each square can contain instructions such as "move one square left," "move one square right," "change what it says in the square," or "stop." In fact, a simple Turing machine will either change a blank square to have a 1 in it, erase the 1 that is there, or leave the square the way it is. It can be shown that such a simple machine can compute all of the kinds of things that mathematicians (or more complex computers) can compute by their usual methods. It is the class of computable (on a Turing machine) methods that will be intended in the following discussions. Such methods are also called *recursive*.

For example, one of the startling results of mathematical logic shortly after Gödel's theorem was published is Church's theorem of 1936:

There is no way that always works to tell whether or not a statement is provable.

Interpreted in light of the Church-Turing thesis, the theorem really means "There is no recursive way that always works to tell whether or not a statement is provable (but we believe that the only possible ways to tell whether or not a statement is provable are recursive)."

Church's theorem is an example of the solution of a *decision problem*. In general, a decision problem always has either the answer "yes" or the answer "no." For example, the decision problem for solving first-degree equations in one variable is "If a, b, and c are numbers and x is a variable, does $ax + b = c$ have a solution?" In this case, you know that the answer is "Yes (unless $a = 0$)." Church's theorem is the answer to the question "If you are given a statement in a system of logic, is there some way that you can tell whether or not the statement can be proved?" Church's answer is "No (using recursive methods of proof)." This does not mean, of course, that there are no statements that can be proved. It simply means that there is no general method of proof that works for all statements.

The mathematician Paul Cohen reports that "it has been said that a famous mathematician once briefly believed that he had found a decision procedure, but luckily for the rest of us, he was mistaken." There are *some* advantages in the "no" answer to the decision problem.

The solution of a decision problem goes far beyond simply proving that one result is not obtainable (such as doubling the cube with straightedge and compass). Solving the decision problem for a class of questions is determining whether or not any effective procedure exists for solving the general questions of the class. If such questions cannot be solved, the class of questions is called *undecidable*. Obviously, the problem of whether or not a class of questions is undecidable is of interest to mathematicians, for it makes no sense to look for a general solution when there is none there.

A PROBLEM OF WORDS

A specific example may help. In 1914, Axel Thue proposed a general class of questions for solution. Although the typical question of this class was not a problem in mathematical logic, solution of the class of questions came to be seen eventually as a decision problem. As such, it was solved in 1947 (independently by Emil Post and A.A. Markov). The problem has come to be known as *the word problem for semigroups*.

(While it is not necessary for you to know what a semigroup is, you might be curious. It is any set plus an operation $*$ so that if a and b are members of the set, then $a * b = c$ implies that c is a member of the set; also, $(a * b) * c = a * (b * c)$. In other words, the operation is closed on the set and associative.)

Thue's notion of the word problem may have come to him from contemplating puzzles such as "Change one letter in "Quite" to get a Harvard philosopher," for the basic notion is transforming one word to another according to certain rules. As mathematicians do, he generalized the process.

Instead of starting right out with a full generalization, however, it is clearer to begin with a partial one. You know that in English there are words that have the same meaning—synonyms—such as push and shove. Furthermore, there are dictionaries of synonyms that pair words that have the same meaning. In such a dictionary, if you looked up push you would find shove listed. Assume that the dictionary is so simpleminded that it has one and only one synonym for each word. However, if "B" is given as the synonym for "A," it may happen that "A" will not be given as the synonym for "B." For example, if you looked up shove you might find thrust instead of push.

There is an operation that may be called Thueing. If you are given a word that has, somewhere in it, a word that is in the dictionary of synonyms, you can Thue the original word by replacing the part with its synonym. If "B" is given as the synonym for "A" and "C" is given as the synonym for "B," then "B" can replace "A" or "A" can replace "B," but in a single Thueing operation you cannot replace "A" with "C." For example, to Thue the word lexicographer you can note that cog is a word in your dictionary of synonyms and that its synonym is tooth. Then, the result of Thueing would be lexitoothrapher. Then the words lexicographer and lexitoothrapher are considered immediately equivalent. A more familiar example of this kind of equivalence is the feminist predilection for replacing the word man in words with the word person, resulting in such locutions as chairperson and personhole.

This notion can be extended by saying that if you have a sequence of words that are all immediately equivalent, then all of the words in the sequence (but especially the first and last) are equivalent. For example, if you focus on the word *graph* in *lexicographer*, you can find definitions in your dictionary of synonyms that might be

graph:	diagram
diagram:	drawing
drawing:	sketch
sketch:	draft
draft:	compose
compose:	set
set:	decree

Using these definitions to Thue *lexicographer* would produce

lexicodiagramer
lexicodrawinger
lexicosketcher
lexicocomposeer
lexicoseter
lexicodecreeer

Therefore, *lexicographer* is equivalent to *lexicodecreeer*.

Now the word problem can be stated: If you are given any specific dictionary of synonyms (like the one described) and any two words, can you tell whether or not they are equivalent?

The word problem for semigroups is just a generalization of the word problem stated in the last paragraph. Specifically, instead of an English alphabet of 26 letters, assume an alphabet of any finite number of letters. In addition to normal words composed of a finite number of letters, also use the *empty word*, the word with no letters. Using this apparatus, it is not difficult to describe exactly what is meant by Thueing and equivalence. Then the word problem for semigroups becomes: Given any finite alphabet and finite synonym dictionary, is there a (recursive) way to determine whether or not any two words are equivalent?

No.

In other words, the word problem for semigroups is unsolvable. However, just as Church's theorem means that not everything can be proved (but some things can be proved), the unsolvability of the word problem for semigroups means that not every pair of words can be determined to be equivalent or not (but some pairs can).

A CONTINUING PROBLEM

The limits of mathematical thought became gradually clearer as the twentieth century progressed. Gödel had shown that there were true statements that could not be proved. Church had shown that there was no effective way of deciding whether or not a particular statement could be proved or not. Now these results might or might not apply to something like Goldbach's conjecture. Still to be exhibited was a specific statement that could not be proved on the basis of some reasonably powerful axiomatic system.

A good candidate for such a specific statement was at hand, however. The problem of proving or disproving Cantor's continuum hypothesis was the first part of the first problem posed by David Hilbert in his famous "shopping list" of unsolved problems in 1900. The continuum hypothesis, however, was resisting all efforts to prove or disprove it. Recall that Cantor had guessed that the second ordinal infinity was the same number as the number of real numbers, a guess that came to be known as the continuum hypothesis. In their efforts to prove the continuum hypothesis, mathematicians found various hypotheses that were logically equivalent to Cantor's version. For example, they showed that if there was no infinity between the number of natural numbers and the number of real numbers, then the continuum hypothesis was true. In spite of reformulating the problem in several ways, however, no one could solve it. Also, as mathematicians will, they generalized the problem, but that did not help either.

Since the continuum hypothesis is a statement about sets, the obvious move was to try to derive a proof or disproof of it from the axioms of axiomatic set theory. Finally, in 1940, Kurt Gödel managed to show that if the other axioms of axiomatic set theory were consistent, then adding the continuum hypothesis to the axioms would not produce a contradiction.

This was insufficient, however, to show that the continuum hypothesis could be proved. First of all, it might be one of those unprovable results that are true anyway, which Gödel himself had found must exist. Furthermore, mathematicians were already well aware of an axiom (usually called a *postulate* in this context) that could be added to a set of other axioms that was not derivable from the other axioms. This axiom is Euclid's fifth (or parallel) postulate.

After many unsuccessful efforts to derive the fifth postulate from the others, mathematicians early in the nineteenth century had found that they could build geometries that denied the postulate. By the end of the century, they were able to show that if geometry was consistent with the fifth postulate included, it was also consistent with suitable denials of the axiom included as replacements for the fifth postulate. Therefore, there was no way to derive the fifth postulate from the other axioms of geometry.

Geometries that replace the fifth postulate with a suitable denial are called *non-Euclidean*. By 1940, mathematicians were generally familiar with non-Euclidean geometries and quite comfortable with the idea. In fact, a non-Euclidean geometry had found a useful role in Einstein's theory of general relativity. So the idea that an axiom could be consistent with the others in a set of axioms but not be provable from the others was well known in geometry and could be seen as a distinct possibility with regard to the continuum hypothesis as well.

Gödel had dropped one shoe in 1940. It was not until 1963 that Paul Cohen dropped the other.

In 1963, Cohen proved that the negation of the continuum hypothesis was also consistent with the axioms of set theory. (He showed, however, that the generalized version of the continuum hypothesis together with the other axioms could be used to prove the *axiom of choice*, a subject that will be discussed more thoroughly in the next chapter.) So here was a statement, the continuum hypothesis, that was shown to be not provable from a generally accepted axiomatic system. In effect, the discovery of the undecidability of the continuum hypothesis said: "Look, these results of Gödel, Church, and others are not just theoretical possibilities; there do exist real theorms in mathematics that can neither be proved nor disproved."

Like Church's discovery that no effective decision procedure exists for symbolic logic (there is no universal way to tell whether or not something is a theorem), Cohen's discovery could be viewed either optimistically or pessimistically.   The pessimist could note that Cohen's work showed that mathematics was not capable of solving certain major questions as formulated.   The optimist could see new horizons available—for example, one branch of set theory with the continuum hypothesis and another without it.   In the optimistic view, set theory had reached the state the geometry had reached a hundred and fifty years earlier.   Now there could be "non-Euclidean" (or perhaps "non-Cantorian") set theory as well as the ordinary kind.

But a central question to the discoverers of non-Euclidean geometry was *which* of the various non-Euclidean or Euclidean geometries described the real world.   Gauss, for one, tried to use data he had collected in surveying to establish whether or not the geometry of space was Euclidean. (The data were not sufficiently precise).   Should there not be a similar effort to find out whether set theory, on which most of modern mathematics is based, is "Cantorian" or "non-Cantorian?"

How would you tell?   Gauss tried to measure the angles in a very large triangle to find out what kind of mathematics is in real space. But how would you measure the two kinds of infinity to find out what kind of set theory God uses?

So far at least, it is not a serious difficulty not to know how to deal effectively with the continuum hypothesis.   No other important ideas in mathematics are dependent upon it.   No engineer would design a machine one way if the continuum hypothesis is true and another way if it is false.   (In fact, if mathematics is merely the formal manipulation of symbols, truth and falsity in the ordinary sense have nothing to with it.)   No one is going hungry because there are ideas that cannot be proved one way or another, and no sick person would become well if someone found an infinity lurking between the number of natural numbers and the number of real numbers.

Few people other than some mathematicians and philosophers have taken seriously the results discussed in this chapter, in fact. Among the few who have, the question has been raised as to whether or not these results indicate something limited about the human mind.   Their assumption is that mathematics is a construct of the human mind.   The limitations perceived in the construct must reflect limitations in the constructor.   Others disagree with this notion very much. That axiomatic systems have limitations need not imply any-thing about the way the mind works, for there is little evidence that human minds, even mathematicians' minds, use axiomatic systems to create new ideas. Instead, the evidence is fairly good that new ideas come from looking at specific examples (nonmathematical induction), from a sense of form (it just feels right), and from sources deep inside that are poorly understood. The Indian mathematician Srini-vasa Ramanujan claimed that a Hindu goddess passed along his ideas to him in dreams. (If so, she often made mistakes.) Maria Agnesi claimed to do some of her best work while sleepwalking. Poincaré worried over one problem for months, getting nowhere. Then, while he was thinking about something else and, in fact, getting on a bus, the answer popped into his mind between the step up and paying his fare. No one has reported achieving really significant results simply by drawing them out of an axiomatic system.

Gödel's theorem, which put the first real chain on axiomatics, was published in 1931. It did not begin to penetrate the public conscious-ness (in the United States at least) until the late 1950s, a quarter of a century later. In the past few years, however, a number of popular accounts of Gödel, Church, and others have been published, some of which have reached a wide audience. No doubt many of those en-countering the limits of axiomatics for the first time will believe that these are also the limits of thought. It is vital, however, to under-stand these results in the context in which they were proved. Each theorem is built on particular axiom systems (or types of systems). There is no free-floating theorem that says:

We cannot learn some things about the world.

In fact, it is known that there are some things about the real world that cannot be found out, but this news comes from physics, not mathematics. The news from the mathematician is that mathematics by itself cannot prove everything about itself. It is truly remarkable that these limitations can be demonstrated, but it should not have been such a surprise that they existed.

7
Misunderstanding Space and Time

It cannot be that axioms established by argumentation can suffice for the discovery of new works, since the subtlety of nature is greater many times over than the subtlety of argument.

Francis Bacon

While mathematicians have made valiant efforts to eliminate the paradoxes of set theory—Burali-Forti's paradox, Cantor's paradox, and Russell's paradox—they have accepted the limitations of axiom systems. In fact, the study of these limitations has become a branch of mathematics with active ongoing investigations of the decision problem for various systems. Most of this work is based to one degree or another on axiomatic set theory, which specifically rules out the known paradoxes—or does it?

Indeed, there are known paradoxes that have not been eliminated in axiomatic set theory and in other branches of mathematics. These mathematical paradoxes are not self-contradictory, like the paradoxes that were discussed in Chapters 5 and 6. Rather, they conflict with intuition. Certainly many of the results of modern physics, which are based largely on mathematical reasoning, are paradoxical in that they also conflict with the normal human experience of the real world. What is to be done with this kind of paradox?

Live with it.

You have already encountered some situations in which the "live with it" decision is generally accepted. Probabilists accept the Petersburg paradox as an unexpected, perhaps unpleasant, real outcome of probability theory. The paradox that a whole infinite set can be matched one-to-one with a part of the set has been more than accepted. It has been embraced. Likewise, the paradoxical behavior of photons and other subatomic particles is the basis for quantum theory, not the disproof of it.

When you become used to these ideas, they do not seem paradoxical any more (although it is hard to get used to the Petersburg paradox). In this chapter, you are given the opportunity to get used to some ideas that are generally accepted, but completely foreign to most people's intuitive understanding of how the world ought to work. The first of these is lodged firmly in axiomatic set theory, which was designed to eliminate paradoxes. Since this one occurs anyway, it is generally accepted as one of life's little surprises, but one that set theorists can live with.

A MODEL PARADOX

When an axiomatic system is put forward in modern mathematics, it is supposed to be purely abstract. Such a formal system involves the manipulation of symbols by specific rules. In a sense, the symbols are not supposed to *stand* for anything. But you know that the mathematicians who developed the system and the ones who use it really have some idea in mind as to what the symbols stand for. Giuseppe Peano developed his axioms as a formal system, but he did so with the whole numbers in mind. The axioms of axiomatic set theory are not supposed to refer to anything in particular, but most mathematicians suspect strongly that they refer to the behavior of collections of objects or ideas—sets, in other words.

It has not gone unnoticed by mathematicians that supposedly formal axiom systems have these interpretations. In fact, the interpretations have been given a name. An interpretation of an axiomatic system is a *model* of the system. The set of whole numbers is a model of Peano's set of axioms. Collections of points in a plane can be used as a model for most of the axioms of set theory, although the number of points in a plane can never be greater than c (the cardinality of the real numbers), and set theory deals with sets that have more members than that.

One of the beauties of axiomatic systems is that they can be constructed to have many models. Results proved with one model in mind will then apply to all of the other models. A set of axioms can leave out many concepts that would specify a particular model in your mind. Consider the following set of axioms for numbers.

1. a. $x + y = y + x$ b. $x \cdot y = y \cdot x$
2. a. $(x + y) + z = x + (y + z)$ b. $(x \cdot y) \cdot z = x \cdot (y \cdot z)$
3. a. $(x + y) \in W$ b. $(x \cdot y) \in W$
4. $x \cdot (y + z) = x \cdot y + x \cdot z$

Since all of these are true for the whole numbers (and since W is the first letter of the word *whole*), it is readily apparent that this system partially characterizes the whole numbers. In other words, the whole numbers are a model for it.

However, the axioms are also true for other sets of numbers. The natural numbers, the rational numbers, and the real numbers are models as well. Even the complex numbers form a model for this system. By adding other axioms to the system, you could reduce the list of models. For example, if you added $x + 0 = x$ as an axiom, the natural numbers would have to be stricken from the list. If you added an axiom that for every x there exists a y such that $x \cdot y = 1$, then the whole numbers would be lost. In that case, you could not interpret W as the set of whole numbers. A different interpretation of W would be required by the other non-whole-number models in any case. Similarly, x, y, and z would have different interpretations for each model. That is, x is a rational number for that model, but a real number for the real-number model.

You can go beyond sets of numbers, however, by also reinterpreting $+$, $\cdot$, $=$, and $\in$. For example, you can interpret x, y, and z as sets. In this model (provided the empty set is included), you can define $+$ as forming the set that has all of the members of x and y (the *union*); you can define $\cdot$ as the set that has all the members that are in both x and y (the *intersection*); and you can say that two sets are equal when they have exactly the same members (ordinary equality for sets). With those definitions, set theory becomes a model for the axiom system.

Or, you can interpret x, y, and z as statements which can be true or false. In this model, $+$ means that $x + y$ is true if either x is true, y is true or both (and false if both x and y are false). Similarly, $x \cdot y$ is true if both x and y are true, but false if either is false. The sign $=$ means that the statement(s) on one side of the sign have the same truth value as the statement(s) on the other side of the sign. In this model, Axiom 4 would be interpreted as "Statement x *and* (statement y *or* statement z) has the same truth value as (statement x *and* statement y) *or* (statement x *and* statement z)."

Models can also be found outside of mathematics. For example, let x, y, and z be interpreted as switches in an electric circuit. As in the previous model, where sentences had one or the other of the truth values *true* and *false*, the switches can have one or the other of the values *open* and *closed*. Let + and · have the same interpretations as in the model just given (for statements), and let = mean that the circuit is in the same state (*open* or *closed*) as indicated by the expressions on either side of the sign. In this model, Axiom 1b is interpreted to mean "For a circuit with two switches, x and y, it does not matter whether x comes first or y comes first, the circuit will be closed only if they are both closed, and it will be open if either is open." You can verify that this and the other axioms are true of real circuits. (It helps to make diagrams.) This last model suggests that it is possible to construct a machine using electric circuits that could perform certain operations in arithmetic or logic.

Use of such deliberately incomplete, or noncategorical, axiom systems is one of the special powers of mathematics. Such systems enable the mathematician to prove many results that will be true in all models that fit the system. A result in number theory will also be true in electric circuits, but only results of number theory that derive from the incomplete axioms.

Gödel's theorem, however, implies that all consistent axiom systems are incomplete. Therefore, even if an axiom system, such as that of axiomatic set theory, is constructed with one particular model in mind, there will be other models that can be invented or discovered to fit the system.

For example, Peano's set of axioms was developed to formalize the set of whole numbers. It is generally agreed that they do this quite well. Recall the interpretations that you were allowed to use for Peano's axioms (in their simplified form):

1. Zero is a whole number.
2. If a number is a whole number, then its successor is also a whole number.
3. a. If two successors are equal, the numbers are equal.
   b. If two numbers are equal, the successors are equal.
4. Zero is not the successor of any number.
5. If zero has a property and if when a number has the property its successor also has the property, every whole number has the property (via mathematical induction).

Since this axiom system must be incomplete, you can find other interpretations for it, models other than the whole numbers.

One model would be to let everything stay the same as in the intended (whole number) interpretation except for S, which would be interpreted as adding 2 instead of adding 1, and W, which would be interpreted to be the set of even whole numbers (even numbers, for short). In that case, Axiom 1, or ($0 \in W$), would become "Zero is an even number;" Axiom 2, or ($a \in W$) → ($Sa \in W$), would become "If a number is an even number, then the number formed by adding 2 to it is also an even number," and so forth. This model is perfectly satisfactory. It is not very enlightening, however, for in some essential way the set of even numbers is just like the set of whole numbers. For example, if you think of the whole numbers as the model of sets of stones (that might be used for counting sheep, for example), then you always get correct results if you take two stones for each sheep instead of one. You can think of the pair of stones as one object. So, in general, the properties of the set of even numbers could be expected to be the same as the properties of whole numbers.

There are, however, models that fit the Peano axioms that are surprising in the way that the even number model is not. For example, consider the numbers in the sequence that can be defined as follows.

Start with the real line. Include between 0 and 1 only those points defined by the sequence whose general term is $1 - (1/n)$. You have the points indicated below

0	$\frac{1}{2}$	$\frac{2}{3}$	$\frac{3}{4}$	...

Now add the points of the sequence whose general term is $1 + (1/n)$

0	$\frac{1}{2}$	$\frac{2}{3}$	$\frac{3}{4}$	...	$1\frac{1}{4}$	$1\frac{1}{3}$	$1\frac{1}{2}$	2

Finally, throw in the points whose general term is $3 - (1/n)$, n greater than 1 (to eliminate a duplication of 2).

0	$\frac{1}{2}$	$\frac{2}{3}$	$\frac{3}{4}$	...	$1\frac{1}{4}$	$1\frac{1}{3}$	$1\frac{1}{2}$	2	$2\frac{1}{2}$	$2\frac{2}{3}$	$2\frac{3}{4}$ ...

This procedure results in an infinite set of points that is clearly unlike the set of whole numbers. For one thing, the whole numbers proceed in an orderly fashion, but this set sort of squeezes down around 1 (which is not in the set), and then proceeds. Also, there are numbers in the set (for example, 2) that are preceded by an infinite number of members of the set. There is no whole number that is preceded by an infinite number of whole numbers (using the normal arrangement by size).

If you replace the set of whole numbers with this set (call it S) and make the meaning of successor be "the next greater number in S," however, it is easy to check that the five axioms of Peano can be modeled by this set. Furthermore, if = is interpreted to mean "is less than," other axioms that characterize the whole numbers can be adjoined to the Peano axioms (or replace them), and S may continue to be a model for the axioms.

This indicates that axiom systems can have models that are not much like the idea that the person developing the system had in mind. In fact, it can be shown that axiom systems always have models that are different from what you expect them to be in the following sense of "different from." If an axiom system is designed to characterize a countably infinite model (such as the whole numbers), there are also models with more than a countable infinity of members that fit the axiom system. Also, for every axiom system, there is a model that is either finite or countably infinite. For example, the real numbers, which are more than countably infinite, can be characterized by an axiom system. But those same axioms must also describe a model that has a countable infinity of members.

Such ambiguities are fine for the kind of limited axiom system you looked at earlier in this chapter, one that was deliberately vague so that it could describe many different things. They are, however, disturbing for axiom systems that are supposed to characterize as completely as possible some set of mathematical objects. For example, since there is a countably infinite model for the axiom system of the real numbers, every result that can be proved from the axiom system applies to the model. In a sense, this makes the "extra" real numbers in *that* model of the system unnecessary. (Maybe you don't need to accept the existence of the square root of 2 after all.)

In 1923, Thorlaf Skolem, who helped to prove the results just cited, suggested an unusually perverse application of those results which has come to be known as *Skolem's paradox* (although it is one of those paradoxes that you "live with"). The axiom system for axiomatic set theory includes an axiom designed to allow the existence of infinite sets—specifically, in fact, of infinite sets that are not countable. But there is a countable model for the axioms of axiomatic set theory. This appears to be such a direct contradiction that Skolem once even suggested that it led him to conclude that axiomatic set theory ought to be abandoned.

Since no one has any idea of how to reconstruct axiomatic set theory so that this paradox does not occur, there have been several attempts to explain it away. One notion is that sets may really be uncountably infinite as seen from "outside" the model, but the model itself does not have the apparatus necessary to notice this uncountability. Another idea is that a set may be uncountably infinite in some models and not uncountably infinite in others (although it has not been explained exactly how this can occur). Of these explanations, the first is more generally accepted.

These explanations, however, have a peculiar character that is not often found in mathematics. Not only are they nonconstructive (that is, they do not show how a model can be constructed that exhibits the properties described), but also they do not show the existence of the kind of model they describe. Instead, they say "Since we know that there is a countable model for the axioms of axiomatic set theory, it is possible that it could have this characteristic which would explain the appearance of an uncountable infinity in the countable model." In this way, it is more like an explanation in physics than one in mathematics. For example, a physicist observes a certain kind of radiation and suggests a possible explanation for its source, which he or she cannot observe. In explaining Skolem's paradox, a mathematician observes a certain property of axiomatic set theory and suggests a possible explanation for its source. The source remains essentially unknown in both examples.

A CHOICE AXIOM

The basic theorem on which Skolem's paradox depends uses the axiom of choice in its proof (although there exist proofs that substitute other controversial steps for the axiom of choice). This is the axiom that was proposed by Ernst Zermelo in 1904 as the basis of his proof that every subset of the real numbers can be arranged so as to have a first member. As soon as Zermelo's proof was published, many mathematicians jumped on him for the new axiom. It was not intuitive, it was not constructive, it was a generally bad idea.

Looking around, however, it soon became apparent that Zermelo's unacceptable axiom had been used unconsciously in the proofs of many basic theorems of mathematics. For example, the theorem that was called upon to show that

$$1 - \frac{1}{2} + \frac{1}{3} - \frac{1}{4} + \cdots + \frac{(-1)^{n+1}}{n} + \cdots$$

has a sum depends on the axiom of choice. So do many of Cantor's results from working with sets. For example, Cantor proved that any infinite set contains a countably infinite subset—but he used the axiom of choice. Even Peano's derivation of the real number system from his axioms for whole numbers used the axiom of choice. Now a great many good mathematicians were saying that the axiom was a "no-no."

But what, exactly, is the axiom of choice? Its essence is that you can choose a subset from any set without specifying exactly how to choose the specific elements of the subset.

For example, in proving that the set of real numbers has more members than the set of natural numbers, you began by choosing the first member of an infinite set of real numbers. You could do this because Zermelo had proved that this was possible in 1904 (using the axiom of choice, which was introduced in that paper). But there was not specific method given for finding that number. In the next step of making a list of real numbers, you repeated the process. Again, however, it was not specified how the number would be chosen. The axiom of choice permits you to do this.

When the dust began to settle on the debate over the axiom, mathematicians found that opinions varied widely. Some mathematicians agreed with Zermelo that use of the axiom made sense for all sets. Others thought it made sense for countably infinite sets, but not for uncountable sets. Still others would only use the axiom of choice for finite sets. And the strict constructionists did not want to choose any set at all unless they had a rule for constructing each member of the set.

One solution to the chaotic situation would be to prove that the axiom of choice was really a theorem of axiomatic set theory. If that could be done, then most mathematicians would accept it. Or, perhaps, it could be shown that the axiom could be disproved from the axioms of axiomatic set theory. Then everyone would view it with suspicion, although some mathematicians would probably prefer to abandon axiomatic set theory than to abandon the results that had been proved with the aid of the axiom of choice. In any case, a result either way would help to resolve the problem.

Various attempts were made along these lines, but the first efforts were all flawed in one way or another. Then, in 1940, in the same paper that demonstrated the continuum hypothesis to be consistent with axiomatic set theory, Gödel also proved the axiom of choice to be consistent with the axioms of set theory. Specifically, Gödel showed that if you assume that axiomatic set theory is consistent without the axiom of choice, then it is also consistent with the axiom of choice included. Therefore, there was hope—but not certainty—that the axiom of choice could be a theorem. It was not certain because there was still the possibility that the axiom of choice was independent of the rest of set theory, just as the parallel postulate is independent of the rest of Euclidean geometry. It is also possible that the other axioms of set theory are inconsistent, but nobody likes to think about that.

Twenty-three years later, Paul Cohen proved that the axiom of choice is independent of the other axioms of set theory. As a result, you can have Zermeloian mathematics that accepts the axiom of choice or various non-Zermeloian mathematics that reject it in one way or another. Recall from Chapter 6 that Gödel and Cohen also proved that there is Cantorian mathematics in which the continuum hypothesis is true and non-Cantorian mathematics in which it is denied (which can also be done in more than one way, as Cohen showed).

Take your pick.

Which mathematics you choose to describe the "real world" is not an easy choice. Rejection of the axiom of choice means rejection of important parts of both "classical" mathematics and set theory. Acceptance of the axiom of choice, however, has some peculiar implications of its own, which will be spelled out in the next section.

VERY ODD CIRCLES (AND SPHERES)

The particular examples that you will be looking at are involved with the notion of measurement. It is easy to take measurement of length or area for granted. In fact, however, a close investigation of measurement by mathematicians in the past 100 years has shown it to be very complicated.

For example, it is quite clear that a line segment (part of a continuum) can be measured in theory

---

but that an isolated set of points such as this one

has no measure—or measure zero, to speak mathematically. Yet it is possible to choose an isolated set of points (without using the axiom of choice) that is contained in a short interval, has measure zero, and still has the same number of members as all the real numbers. Intuitively, you would expect that a set of points that had the power of the continuum as its cardinal number would not be squeezed into a short interval unless there were some place in the interval that had a positive measure.

The particular construction used to obtain a set of points with this property and the proof of one-to-one correspondence are most easily done using a nondecimal notation. If you went to school during the "new math" period in mathematics education (roughly between 1960 and 1975), then you probably have had some exposure to other number bases. The decimal system that you use daily is based on 10. That is, '432.156' means

$$4 \cdot 10^2 + 3 \cdot 10^1 + 2 \cdot 10^0 + 1 \cdot 10^{-1} + 5 \cdot 10^{-2} + 6 \cdot 10^{-3}$$

(where 10^{-1} means $1/10^1$, 10^{-2} means $1/10^2$, and so forth). For some purposes, bases other than 10 are useful.

For example, computers usually use some variant of a system based on 2. In that system, only the digits 0 and 1 are used, for the decimal numeral '2' is represented by the base two numeral '10,' which means $1 \cdot 2^1 + 0 \cdot 2^0$. With the use of a point (like the decimal point, but this would have to be called the *binary point*), you can represent all real numbers in base two. For example '1.1_{two}' means the same as '1.5_{ten},' since $1.1_{two} = 1 \cdot 2^0 + 1 \cdot 2^{-1}$, or $1 + (1/2)$. Some rational numbers require repeating numerals. Just as $1/3$ in base ten is shown by the repeating decimal $0.333 \cdots$, or $0.\overline{3}$, the base two equivalent, $1/11_{two}$, also repeats, becoming the numeral $0.010101 \cdots$, or $0.\overline{01}$. Interestingly, however, numerals for rational numbers that repeat in one base may not repeat in some other base. For example, $1/3$ is not a repeater in base three.

For irrational real numbers in base two, you also must use non-repeating, nonterminating numerals. For example, $\sqrt{2_{ten}}$ is the same as $\sqrt{10_{two}}$ and its representation with the binary point is $1.0110101 \cdots_{two}$, which means the same as

$$1 \cdot 2^0 + 0 \cdot 2^{-1} + 1 \cdot 2^{-2} + 1 \cdot 2^{-3} + 0 \cdot 2^{-4} + 1 \cdot 2^{-5}$$
$$+ 0 \cdot 2^{-6} + 1 \cdot 2^{-7} + \cdots$$

or

$$1 + \frac{1}{4} + \frac{1}{8} + \frac{1}{32} + \frac{1}{128} + \cdots$$

What you need in the following work, however, is not base two (although that is used at the end), but base three. The principle is exactly the same. There are now three digits that can be used: 0, 1, and 2. A numeral such as '210.102_{three}' means

$$2 \cdot 3^2 + 1 \cdot 3^1 + 0 \cdot 3^0 + 1 \cdot 3^{-1} + 0 \cdot 3^{-2} + 2 \cdot 3^{-3}$$

or

$$2 \cdot 9 + 1 \cdot 3 + 0 \cdot 1 + 1 \cdot \frac{1}{3} + 0 \cdot \frac{1}{9} + 2 \cdot \frac{1}{27}$$

or $18 + 3 + (1/3) + (2/27)$, which in ordinary decimal notation is $21 \text{-} 11/27$. The base three system can also be used to represent all real numbers.

Now you are ready to dissect a short line segment. Start with the segment from 0 to 1 including both endpoints.

0   1

Now remove all of the points from 1/3 to 2/3, keeping only 2/3

0 $\frac{2}{3}$ 1

In base three notation, you have removed all points on the line that correspond with the real numbers that are represented in base three with a 1 after the base three point. That is, every number that starts out $0.1\cdots$ is in the set removed, and every number that starts out either $0.0\cdots$ or $0.2\cdots$ is in the set kept. (Just as 1_{ten} can also be represented as $0.\overline{9}_{ten}$, 1_{three} can also be represented as $0.\overline{2}_{three}$.)

Now remove from what is left of the line the middle third of each of the two pieces. As before, remove the left endpoint but keep the right endpoint.

0        $\frac{2}{9}$ $\frac{2}{3}$ $\frac{8}{9}$ 1

In terms of base three numerals, you have now removed every point that matches a real number that has a '1' in the second place after the base three point. In other words, since you have already removed the numbers that start out $0.1\cdots_{three}$, this step removes the numbers that start out either $0.01\cdots_{three}$ or $0.21\cdots_{three}$.

Continue the process of removing the middle third of each segment a countable infinity of times. Always remove the left endpoint, but keep the right endpoint. The result will be a line that is not a continuum. Between any two points on the line, there will be a part that you have removed. In fact, you will have a set of isolated points.

How many are in the set? You can use base two to find out. At each step in the removal process, when you looked at it in base three, you found that you were removing all and only those numbers that had a 1 showing in a particular place after the base three point. For the first step, you removed all the points that had a 1 in the first place; for the second step, all with a 1 in the second place; for the third, all with a 1 in the third place. So, after a countable infinity of steps, you have removed all points whose representations in base three have a 1 in them in any place. (You could prove this by mathematical induction if you wanted.)

But the points that are left are just those composed of nonterminating sequences of 0's and 2's. They might as well be base two numerals since they use only two digits.

In fact, since the base two numerals for the real numbers include *all* nonterminating sequences of 0's and 1's, you can set up a one-to-one correspondence between the points that are left on the line after the removal process and the base two numerals for all real numbers. Simply match each sequence of 0's and 1's in base two with the corresponding sequence of 0's and 2's in base three, corresponding by substituting a 1 for each 2 that occurs. This proves that the number of points left on the line has the same number as the set of real numbers, the number often called c (for *continuum*).

(This result should not be a total surprise, since Cantor had shown that removal of a countable infinity from an uncountable infinity leaves an uncountable infinity.)

And yet, the set of points left after the removal process has no length. It is of "measure zero," as mathematicians say. When you consider this closely, it is rather astonishing.

It turns out, however, that many intuitive concepts about measurement are a bit shaky. The problems begin to emerge when you start to put measurement on an axiomatic basis. For example, here are some possible axioms that would apply to measurement in general; that is, length, area, and volume (or even measurement in dimensions higher than 3) would be covered by the axioms.

1. For every geometric figure A there is a nonnegative real number $|A|$ called its *measure*, unless its measure is countably infinite. (The number can be 0, which is nonnegative although not positive.)

2. If A and B are geometric figures, and A is congruent to B, then

$|A| = |B|$. (For the purpose of this discussion, you can say that A is congruent to B if there is some way to move B around without changing B in any way so that B and A can occupy the same place; that is, every point of B is now on a matching point of A with no points left over.)

3. If $A_1, A_2, A_3, \cdots, A_n$ are geometric figures such that for any pair A_i and A_j the measure of their intersection is 0, and such that the union of all of them forms a geometric figure A, then
$$|A| = |A_1| + |A_2| + |A_3| + \cdots + |A_n|$$

If you examine these axioms closely, you will find that they correspond pretty well to your experience with length, area, and volume with the exception perhaps of admitting 0 and infinity. But admitting 0 and infinity has the advantage of allowing you to measure any geometric figure whatsoever, so most mathematicians consider the advantage to outweigh any possible disadvantages.

This system quickly leads to some nonintuitive results. For example, one geometric figure can be congruent to a subset of itself (that is, a proper subset—one that does not contain every point in the original figure). And, therefore, a figure can have the same measure as one of its proper subsets. Both of these notions strike one as "wrong" until you look at a specific example.

Consider the following geometric figure, which is a ray that has an endpoint at P and continues without bound beyond the point Q on the ray.

Now place a point R on the ray between P and Q.

The ray RQ is a proper subset of the ray PQ. But the ray RQ is congruent to PQ since you can slide all of the points of RQ back to where R is coincident with P and every point of RQ will be superimposed on a corresponding point of PQ.

Furthermore, as the axioms imply, $|RQ| = |PQ|$. Both are countably infinite. (The set of points on a ray are uncountably infinite *sets*; but the *measures* of these sets are countable.)

You may think this is cheating, since everyone knows that an infinite set can be put into one-to-one correspondence with some of its proper subsets (as was discussed at length in Chapter 5). This is a little different, however, for the idea is that the sets are beyond one-to-one correspondence—they are congruent. A line segment can be made to be in one-to-one correspondence, point for point, with a square; but the two geometric figures can never be congruent.

It is true, however, that the use of geometric figures whose measure is infinite (that is, their measure increases without bound) is cheating in a way. For it can be shown that no two figures of finite length such that one is a proper subset of the other can be congruent. Notice that this result is specified for *length* only.

When you move to area and volume, the situation becomes immensely more complicated. Here are three examples (which are offered without proof, but good mathematicians who have examined these proofs assure us that the proofs are valid).

Schwarz's Paradox

About a hundred years ago, H. A. Schwarz discovered something very peculiar about the area of the curved part of a cylinder, the geometric figure that looks like a tin can. Suppose that you have a cylinder that is one unit high and has a base with a radius of one unit. If you unwrap the curved surface and flatten it out, you will get a rectangle whose height is 1 unit and whose base is 2π units. Therefore, the area of the curved surface must be 2π square units.

Now this rectangle can be broken up into triangles that get smaller and smaller. Here are the first three steps in separating the rectangle into such triangles.

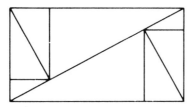

Each of these triangles obeys the conditions of Axiom 3, so the sum of their measures must be the same as the measure of the original rectangle. But Schwarz showed that the limit of the sum as the triangles went to measure 0 (that is, as they became smaller and smaller) was a number that increases without bound—infinity, in other words. So the area of the parts is infinite, but the area of the whole is 2π.

Hausdorff's Paradox

This time consider the surface of a sphere. Around the time of World War I, Felix Hausdorff was investigating measurement of the area of the surface of a sphere, and he reached the following remarkable conclusion.

It is possible to divide the surface of a sphere into three congruent figures, A_1, A_2, and A_3, that meet the conditions of Axiom 3—the union of A_1, A_2, and A_3 is the sphere and the intersection of each pair of them has measure 0—such that each subset of the sphere is congruent to the union of the other two subsets as well.

In other words, A_1 is congruent to A_2 and also to the union of A_2 and A_3; A_2 is congruent to A_3 and also to the union of A_1 and A_3; and A_3 is congruent to A_1 and also to the union of A_1 and A_2. This would not seem so startling if they all had measure 0, but that cannot be, for they make up the entire surface of the sphere, which (for a unit radius) has measure 4π.

The Banach-Tarski Paradox

A few years later, Stefan Banach and Alfred Tarski used ideas similar to Hausdorff's to investigate volume. What they discovered was even more amazing than what Schwarz and Hausdorff had discovered about area.

Banach and Tarski defined a notion that they called *equivalent by finite decomposition* (which, for the purposes of this discussion, will be abbreviated to *equivalent*). Two geometric figures are equivalent if they can be separated into a number of subsets that meet the conditions of Axiom 3 and each subset of the first figure is congruent to one and only one subset of the second. This is a good intuitive notion. It means that if set A is equivalent to set B, then you could use the pieces (subsets) of set B to put together a figure congruent to set A. Similarly, you could use the pieces of set A to put together a figure congruent to set B.

Tangram puzzles are an example of a two-dimensional version of equivalence by finite decomposition. You start with pieces that can be assembled into a square or other shapes without overlapping. The puzzle comes in trying to make a specified shape from the set of pieces.

Geometric decompositions are another example of the same idea. In performing a geometric decomposition, however, you are not given the cuts that separate the first figure into the subsets that they are to be assembled into. The goal in a geometric decomposition is to make as few cuts as possible.

Just to be certain that this is clear, here is a specific geometric decomposition problem. The original geometric figure (call it set A) is three congruent squares. The problem is to use just six cuts to make subsets that can be assembled to form a new square (set B) that is equivalent to set A (by finite decomposition as defined above).

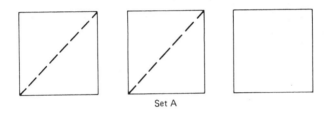

Set A

In the diagram above, the dotted lines indicate the first two cuts to be made. Then group the triangles around the remaining square as shown below.

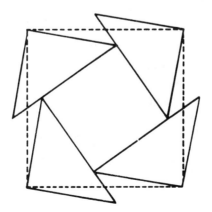

Here, the dotted lines indicate the square that will be set *B*. Now all you have to do is to cut off the parts of the triangles that are outside the dotted lines (four more cuts). These will fit into the dotted lines.

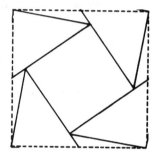

The dotted lines in the third diagram show the edges of the pieces of the original three squares.

This is a real geometric dissection. There are, however, many "fake" geometric dissections that produce fallacies. For example, there is the *Curry triangle.*

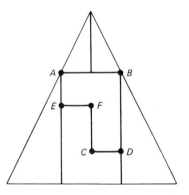

You can make this triangle with a base of 10 centimeters and a height of 12 centimeters. The line segment *AB* is 7 centimeters from the base. The short segment *CD* is 2 centimeters from the base. The segment *EF* is 3 centimeters farther. All the angles that look like a right angle are. Make this on paper that has one colored side, so that you will be able to tell the front from the back. Cut out the pieces and save the trianglar "hole" in the paper.

Now turn the pieces over and rearrange them so.

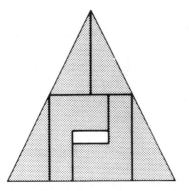

You will discover that this triangle has the same outer dimensions as the original one. But it has a "hole" in it that is two centimeters long and a centimeter high. As Ronald Reagan said, "Where's the rest of me?" (You can check the outer dimensions by laying the "hole" from which the triangle was cut over the new "holy" triangle.)

Of course, this geometric dissection problem is a cheat, like the geometric fallacies in Chapter 3. If you calculate the areas of each of the pieces and add them together, they come to 59 square centimeters. But the original triangle was supposed to have a measure of 60 square centimeters and the "holy" triangle is supposed to have a measure of 58 square centimeters (with the hole subtracted).

A clue to what went wrong is that the triangle formed by the two top pieces of the original triangle is supposed to be similar to the original triangle. It cannot be, however, with the dimensions given. The same problem exists with the "holy" triangle. If the drawing is made sufficiently accurate, then there will be small gaps (one square centimeter's worth) between pieces of the original triangle. Similarly, there will be small overlaps (one square centimeter's worth) for the pieces of the "holy" triangle.

To return to the main subject: What Banach and Tarski did was to investigate similar dissection problems for both volume and the surface of a sphere. What they found was that results similar to the Curry triangle appear, only more so. And they appear without cheating.

The general statement of their result can be paraphrased as follows: Consider what you usually think of as "ordinary three-dimensional space." Any higher-dimensional space that is basically like the usual concept of space is called *Euclidean*. In a Euclidean space, such notions hold as a higher-dimension version of the Pythagorean theorem. This is not the case in higher-dimensional non-Euclidean spaces.

What Banach and Tarski showed is that: If you have two geometric figures in a Euclidean space that consist of an ordinary outside surface and some points inside, then there is a finite decomposition that shows that they are equivalent. (The same situation exists for the surface of a sphere.)

When this result is reported to nonmathematicians, it is often stated something like this: Banach and Tarski proved that it was possible to disassemble a small sphere, such as a pea or a grapefruit, and reassemble it with no interior gaps into a sphere the size of the earth or the sun. Indeed, that is one of the consequences of the theorem.

Another consequence, which is really just as alarming, is that you can take a sphere that has a radius of one unit (a "unit sphere") and cut it into nine pieces that have the following property. Five of the pieces can be put together to form a unit sphere (with no gaps). The remaining four pieces can also be put together to form a unit sphere (with no gaps).

This is really getting something for nothing. Suppose that you had a sphere made of gold. Then all you would have to do to double your wealth, according to Banach and Tarski, is carefully to cut the sphere apart into nine pieces, glue them together properly, and then you would have two. Obviously, the same process could then be repeated with each sphere, giving you four golden spheres. And so forth. What a great way to become rich!

The problem is the axiom of choice. Both the Hausdorff paradox and the Banach-Tarski paradox depend on the axiom of choice. One result of this is that Banach and Tarski do not, alas, give you a method for accomplishing their duplication of the sphere. They simply show that if you can pick one member from each of an infinite number of infinite subsets, then it can be done.

While you might think that a result like the Banach-Tarski paradox would cause mathematicians to switch to non-Zermeloian mathematics, no such thing has happened. Instead, the mathematicians who have thought about it have decided that Banach-Tarski is one of those paradoxes that "you just live with." They would rather abandon the naive concept of volume than abandon the axiom of choice.

There has been little concern about the Banach-Tarski paradox since World War II.

Hausdorff's results suggest that we do not understand area very well. Banach and Tarski showed that we do not understand volume very well. (There is a pretty good understanding of length, however; so all is not lost.) Another way to look at this lack of understanding is to say that we do not understand space very well.

But Banach and Tarski worked in a mathematical space, specifically Euclidean n-space, n greater than 2. There is good reason to believe that the space we inhabit is not Euclidean three-space; that is, the space of the universe is not three-dimensional but four-dimensional, and furthermore the universe does not work the way Euclid's postulates said it would. But for a better understanding of that, and the paradoxes it brings, it is necessary to leave the realms of pure mathematics and once again sojourn with the physicists.

HOW TO STAY YOUNG

One of the best-known facts about Euclidean three-space is that the shortest distance between two points is a straight line. In a sense, however, in the real space you live in, the *longest* distance between two points is a straight line. One result of this is that it is possible for one member of a pair of twins to be considerably older than the other twin.

This paradox occurs because, according to Albert Einstein's theories of relativity, you do not live in Euclidean three-space. You live in Minkowski four-space. Einstein's special theory of relativity, which leads to the Twin paradox, is one of the most thoroughly verified theories of physics (only quantum electrodynamics—with *its* paradoxes—has more experimental verification). While the special theory can be explained without resorting to Minkowski four-space, the concepts of Minkowski four-space make some of its results—including the Twin paradox—easier to understand (for mathematically minded folks anyway).

(It is worth noting that many of the geometric concepts involved in relativity were developed before Einstein was born. These are the concepts of non-Euclidean geometry. Hermann Minkowski quickly saw how these concepts could be applied to relativity theory.)

To begin, here is an easy way to describe the differences between various *n*-spaces, as *n* changes from 1 to infinity. The number *n* is just the number of numbers needed to locate a point in the space. For example, if you know that point *P* is somewhere along this line segment,

$$A \text{————————————} B$$

then you can describe where *P* is by assigning a single number. Suppose that you take the distance from *A* to *B*, $|AB|$, as one unit. Then, if *P* is 0.5 unit from *A*, it is smack in the middle of the segment and no place else. A line segment, when considered as a space, is one-space, since it takes one number to locate a point. Notice that the same would be true if the line segment were replaced by a curve from *A* to *B*.

As long as you know the distance from *A* to *P*, measured along the curve, you can locate *P* on the curve.

Now consider a square region. A square region is two-dimensional because it takes two numbers to locate a point in a square region.

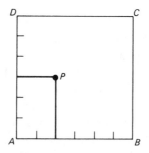

In this case, *P* is two units from *AD* and three units from *AB*. The region is a two-space.

Similarly, it takes three numbers to locate a point in a cubic region. Since it also took three numbers to locate a point anywhere in the prerelativity universe, it was assumed that physics could be encompassed by three-space (although after the discovery of non-Euclidean geometries, it was not clear that the three-space of the universe would turn out to be Euclidean; this was a separate question from that of the dimension of the universe).

In 1905, however, Einstein showed that it takes four numbers to locate a point in the universe.

This is a direct consequence of the paradoxical fact (proved in many well-known experiments) that light always travels the same speed in a vacuum. In fact, if two persons are measuring the speed of light from a single source, they get the same result no matter what their relative motions with respect to the source are. If one is moving rapidly toward the source and the other is moving rapidly away from it, it does not matter. The speed of light stays the same.

This constant speed of light in a vacuum, about 300,000 kilometers per second, is called c by physicists. Because mathematicians use c for the power of the continuum, this could cause some confusion. You can avoid this and also simply the mathematics of what follows by choosing your units of length and time so that c is equal to 1. That is, one unit of distance will be 300,000 kilometers, one unit of time will be 1 second, and the speed of light in a vacuum will be 1 unit of length per unit of time.

In addition to always being the same, the speed of light has another remarkable property. It is as fast as any material thing can travel. (For the purpose of this discussion, it is not necessary to explain why that is so. Take my word for it, or read any book on relativity theory.) Consequently, information reaching you by light waves is the fastest way that you can get information about what is happening anywhere that you are not.

Now consider the following situation. A person on the surface of the earth sees, at some distance away, two signals, also on the surface of the earth, that go off at the same time. Another person, flying swiftly in a jet plane, also sees the two signals. However, in the time that it takes light to travel from signal A to the plane, the plane will have moved. Say that the plane is moving along a path parallel to the line from A to another signal, B, heading away from A toward B. While the light is traveling from A, the jet plane is moving away. It takes longer for the light from A to reach the plane than it does for the light from B, for the jet plane is getting closer to B. Consequently, the person on the jet plane does not see the signals go off at the same time.

For the normal speed of a jet plane, the difference in time would not be noticeable. But if the airplane were somehow traveling near the speed of light, the difference in time would be quite noticeable.

From the point of view of the person on earth, the *time* of the two signals was the same. From the point of view of the person in the plane, the *time* of the two signals was different. When you begin to unravel the complications of this, as Einstein did, you discover that time really is different for the person on the plane from time for the person on earth. In fact, to locate a point in the universe you need four numbers. The first three numbers give the position with respect to a particular cube and the other number is needed because the point may be moving with respect to the cube. The fourth number is time. Time is different for a point moving with respect to the cube than it is for a point that is stationary with respect to the cube.

When Minkowski applied the results of relativity theory to the development of a four-dimensional geometry of space-time, he found that it was not a Euclidean four-space. Although Minkowski space resembles Euclidean space in many ways, the concept of length is quite different.

Working with four-space is difficult to imagine and impossible to draw in any way that is very helpful. On the other hand, all of the essential concepts are the same in two-space, which is easy to show on paper. Therefore, instead of comparing Minkowski four-space with Euclidean four-space, it is instructive to compare Minkowski two-space with Euclidean two-space.

First of all, consider length in Euclidean two-space. For a straight line segment, length is the same as the distance between two points. (The same ideas generalize to curves that are not straight, but involve mathematics that is beyond the scope of this book.) You learned how to find the distance between two points in Euclidean two-space in high school algebra or geometry.

On the next page are two points located in Euclidean two-space. They are the endpoints of a line segment. The square of the distance between them can be found by the Pythagorean theorem:

$$d^2 = (x_2 - x_1)^2 + (y_2 - y_1)^2$$

Then $\sqrt{d^2}$ is the length of the line segment.

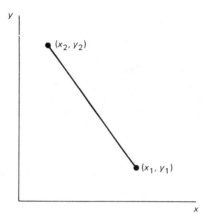

For example, if (x_1, y_1) is $(4, 1)$ and (x_2, y_2) is $(1, 5)$, then the square of the distance between the points is $(1 - 4)^2 + (5 - 1)^2$, or $(-3)^2 + (4)^2 = 9 + 16 = 25$. Since $d^2 = 25, d = 5$.

Now replace the y axis with an axis for time, t. If time were the same for two observers moving with respect to each other, the space would be Euclidean and the same distance formula would apply with just t substituted for y. You know, however, that time changes for a moving observer. The effect of this on the graph is that the moving observer gets different coordinates for (x, t) than the stationary observer does. We cannot go into the mathematics in detail, as the problem is slightly beyond our resources. The change, however, turns out to be the same as would occur from a rotation of the axes.

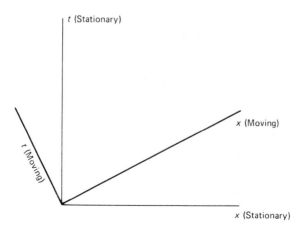

One way to indicate such a rotation is to multiply the time axis by a multiple of i, the imaginary number that is the replacement for $\sqrt{-1}$. Since $c = 1$ (the speed of light in a vacuum is 1) in this discussion, the multiple, which would be ic, becomes simply i.

Therefore, the distance formula is modified by replacing y with it. In this system, then,

$$d^2 = (x_2 - x_1)^2 + (it_2 - it_1)^2$$

You can, however, factor $i^2 = -1$ from the time part of the formula, which gives

$$d^2 = (x_2 - x_1)^2 + (-1)(t_2 - t_1)^2$$

or

$$d^2 = (x_2 - x_1)^2 - (t_2 - t_1)^2$$

Now look at the same two points in Minkowski two-space. Instead of the y axis, the vertical axis is a t (for time) axis. The distance between the points is defined to be $d^2 = (x_2 - x_1)^2 - (t_2 - t_1)^2$. The length of the line segment is still $\sqrt{d^2}$. However, if (x_1, t_1) is $(4, 1)$ and (x_2, t_2) is $(1, 5)$, then the square of the distance between the points is $(-3)^2 - (4)^2 = 9 - 16 = -7$. This means that the length of the segment is $\sqrt{-7}$, or $i\sqrt{7}$. In other words, the length is given by an imaginary number.

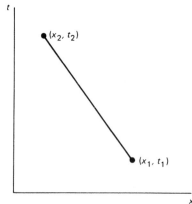

This formula was derived from experiments. There is no way you can determine whether or not you are in Euclidean space or Minkowski space by reasoning alone.

Now apply the formula for distance in Minkowski space to an interpretation of motion. Suppose that the point P is one unit away from the origin of the x axis. If it is not moving along the x axis, it is still moving through time. Therefore, its graph is a vertical line. The line shows that the point is making no motion along the x axis, but is moving continuously through time.

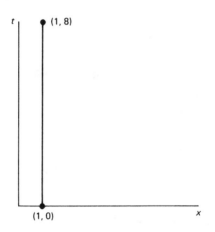

Say that the point moves through time to where $t = 8$. That is, the point does not move along the x axis for eight seconds. Then the square of the length of its graph, or *world line* as it is called in relativity theory, is $(1 - 1)^2 - (8 - 0)^2 = -64$. Therefore, the length of the world line is $8i$.

Now consider another point. This one is *moving* along the x line. Say that the point moves from $x = 1$ in a constant motion to $x = 4$, then back to $x = 1$ at the same speed (but in the reverse direction). This motion also takes 8 seconds. The graph below shows the world line of this point (as it moves through time) as well as the world line of the point that you considered before, the one that stayed at $x = 1$ and just moved through time. Notice that both points move through time, reaching the point $(1, 8)$.

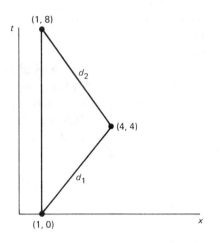

The big difference is the length of the paths, or world lines, they took to get to (1, 8). You know that the length of the path of the first point was $8i$. Now calculate the world line of the point that moves along the x axis. The distance from (1, 0) to (4, 4) can be found using

$$d_1^2 = (4 - 1)^2 - (4 - 0)^2$$
$$= 9 - 16 = -7$$

Therefore, $d_1 = i\sqrt{7}$. Similarly,

$$d_2^2 = (1 - 4)^2 - (8 - 4)^2$$
$$= 9 - 16 = -7$$

So d_2 is also $i\sqrt{7}$. Then the total length of the world line of the moving point is $2i\sqrt{7}$. The square root of 7 is about 2.646, so $2i\sqrt{7}$ is about $5.392i$.

As predicted earlier, the straight world line between two points is longer than another world line between the two points, for $8i$ represents a longer world line than $5.392i$ does. The physical effect that corresponds to this geometric situation is that time is slower for the point that moves along the x axis than it is for the point that stays put.

In fact, you can see from the graph that the time of a point that is stationary with regard to the x axis is the fastest time there can be. Picture the point moving through time along its world line. Any deviation from the vertical line will cause a slowing down of time. And the only way that the point can deviate from the vertical is if it moves with respect to the x axis.

The result is the *twin paradox*. If one twin stays at home and the other roams about, when they get together again the roaming twin will be younger than the "stay-at-home." Of course, the roaming has to be done at speeds that are fairly close to the speed of light for the difference in their ages to be noticeable. For example, if the roaming twin travels to the nebula Andromeda, which is two million light-years away, at speeds that get him there and back in three million years, he or she will return home only 29 years older. That is, three million years will have passed on earth, while only 29 years will have passed in the spaceship.

To take a more moderate example, if an astronaut made a trip of one year at an average velocity of half the speed of light, when he returned home a year and nearly two months would have passed. On a ten-year trip at the same average speed, the twin astronaut would return to earth more than a year younger than the "stay-at-home" twin.

Several objections to the paradox should come to mind. For one thing, the earth is not a stationary point. It is moving around the sun, the sun is moving around the galaxy, and the galaxy is going somewhere too. But, what is the set of axes with regard to which all this motion is taking place? One consequence of the speed of light staying the same for all observers, no matter how they are moving, is that you cannot deduce any preferred set of axes. It is entirely reasonable to choose the axes as stationary with respect to the earth. Then everything else is moving, just as the ancients perceived. The astronaut's motion is then taken with respect to the axes that are stationary for the twin on earth.

Hold on! If there is no preferred set of axes, why not take the set that is stationary with respect to the astronaut? Then the astronaut twin will not move, but the "stay-at-home" twin will travel away from the astronaut twin at speeds averaging half the speed of light, taking the whole earth with him. After ten years, when the "stay-at-home" brings the earth back to the stationary astronaut, the "stay-at-home" will be more than a year younger than the astronaut, the reverse of what happened when the axes were kept stationary with respect to the earth.

People have tried to use the argument of the last paragraph to refute the twin paradox. Clearly, it is not possible that the same twin could be both older and younger than the other twin. If you have the freedom to choose any axes you want, then why does the twin paradox still hold true (which it does; experimental evidence is very strong in its favor)?

The problem is resolved as follows. Take the earth as not accelerating in space, so that a set of axes stationary with regard to the earth is the stable reference set. The astronaut cannot travel away from the earth and come back to it without accelerating somewhere along the line. Even if the astronaut were already traveling at a constant speed as he or she passed earth, the turnaround to come back to earth would involve acceleration. A set of axes that is accelerated is not the same as one that moves with a constant speed.

Einstein's other theory of relativity, the general theory, helps make this point clearer (although the twin paradox would be a reality even if the general theory were not true). In the general theory, Einstein points out that passengers in a sealed rocket ship with no windows would have no way of telling the forces caused by acceleration from the force of gravity. He concluded from this and other evidence that these forces are *exactly* the same. One consequence of the general theory is that this force, gravity or the force caused by acceleration, also slows down time. From that point of view, it is the acceleration that *causes* the traveling twin to return younger than the "stay-at-home."

(Incidentally, the general theory also shows why it is a good idea to have restaurants on the tops of tall buildings. Because gravity is slightly less at the top of a tall building, time there moves slightly faster than on the surface of the earth. A meal at Windows on the World, atop the north tower of the World Trade Center, takes less time than a comparable meal on the ground.)

The twin paradox, like the Banach–Tarski paradox, points up the conclusion that naive ideas of space and time are incorrect. About 2500 years ago, using axes stationary with respect to the earth, the Greek philosopher Zeno of Elea also tried to show how naive ideas of space and time led to paradoxes. In the last chapter, you can see how his ideas have fared since then.

8
Moving Against Infinity

Experience has taught most mathematicians that much that looks solid and satisfactory to one mathematical generation stands a fair chance of dissolving into cobwebs under the steadier scrutiny of the next.

Eric Templeton Bell

Sooner or later, everyone who writes about paradoxes must deal with the paradoxes that were proposed sometime around 450 B.C. by Zeno of Elea. Dealing with this subject is complicated because practically everything anyone knows about Zeno is secondhand. While Zeno probably proposed about 40 paradoxes, the four best known are found only in a discussion of them by Aristotle (and then in innumerable discussions of them by later writers, based upon Aristotle). Two or three others have come down from other sources, perhaps in Zeno's original words. Thus, it is not fully clear what Zeno had in mind in proposing his paradoxes. What is clear is that the problems they introduced into mathematics upset the Greeks immediately and continue to be controversial to this day.

It is generally believed, however, that Zeno was trying to undermine the philosophy of the Pythagoreans by a form of indirect proof. In other words, he took the assumptions of Pythagorean mathematics and philosophy (which for Pythagoreans were pretty much indistinguishable) and showed that these assumptions led to contradictions.

From the evidence of the known paradoxes, he did something slightly more complicated than that. In some paradoxes, Zeno assumes that the real world is made up of elements that correspond to mathematical points—locations with no size. He then shows that this assumption leads to contradictions with what people perceive in ordinary life. In other paradoxes, Zeno begins with the opposite point of view that the real world is made up of indivisible bodies that have the characteristics of normal objects, but on a small scale. You can think of these as *Democritus' atoms*. With this assumption he also shows that you can derive contradictions with what people perceive around them.

Thus, if the argument was between those who thought of the real world as infinitely divisible and those who thought of the real world as composed of "atoms," Zeno's point of view seems to be "A pox on both your houses."

Ignoring the philosphy for a moment and looking at what people do in practice, there appears to be a division today between the two points of view that Zeno attacked. Most mathematicians proceed about their business *as if* the real world were infinitely divisible. And they obtain correct results that apply to the real world from this assumption. Physicists, however, operate on the notion that the real world is made out of chunks called *quanta* that behave the same way that Zeno's indivisible bodies do. And *they* obtain correct results that apply to the real world from this assumption.

You can probably deduce that mathematicians have been more concerned with Zeno's arguments against the idea that the real world is composed of mathematical points. After all, in those paradoxes Zeno is attacking mathematics. Consequently, there are many discussions of what those paradoxes mean, and they have been disposed of over and over. Nevertheless, the people who have thought most critically about the foundations of mathematics do not accept the arguments of those who have proclaimed these paradoxes resolved.

The notion that the real world is composed of something like mathematical points can be called the *idea of the continuum*. The word *continuum* is familiar as the name used by Cantor and other mathematicians to refer to the real line. In general, it appears that if you understand the meaning of the real line, the other questions that might emerge concerning continuity can be solved.

The essential problems to solve are those concerned with limits and continuity. When Newton and Leibniz developed the calculus, their explanations of why their new methods worked did not solve these problems. Consequently, their ideas were attacked for being full of paradoxes. A major problem was that a quantity was very close to zero, but not zero, during the first part of the operation; then it became zero at the end. When you "took the limit," as folks say today, quantities that existed before suddenly ceased to exist. One critic, Bishop George Berkeley, called the result the "ghosts of departed quantities."

In the nineteenth century, A. L. Cauchy and Karl Weierstrass resolved these paradoxes by defining them away. That is, the definitions of *limit* and *continuity* were given in such a way that the startling change from not-zero to zero was avoided. In doing so, they came down firmly on the side of a point having no size—just a location—in mathematics. Furthermore, by identifying points and numbers as the same thing mathematically, geometric intuition could be eliminated, since it was possible that the geometric concept of a point could mislead the mathematician. In fact, Weierstrass and others defined curves or sets of points that defied geometric intuition. Numbers were safer.

For example, curves were shown to exist that were continuous, but which had no point along them at which one could draw a tangent. Peano and others found curves that completely filled two-dimensional regions. These kinds of anomalies had been tacitly assumed not possible before.

Near the end of the nineteenth century, Richard Dedekind proposed changing the definition of real number to avoid any possible influence from geometric intuition. Instead of saying that a real number is the coordinate of any point on the continuum (or an infinite decimal, which can be shown to be equivalent to being a coordinate), Dedekind defined the real numbers as infinite sets of rational numbers. The notion was that people intuitively understood the natural numbers and the whole numbers; besides, these number systems could be put on an axiomatic basis (of course, this was long before Gödel). From the whole numbers it is easy to make purely mathematical definitions for the positive and negative integers and for the rational numbers. Dedekind's definition, then, built upon all these well-understood systems.

Dedekind based his work on completed infinite sets. Within ten years of his proposal, the paradoxes of infinite sets began to be bruited about in mathematical circles. Nevertheless, most mathematicians today feel that Dedekind's definition is the basic one that describes what a real number "really" is.

When they are trying to solve a problem, however, mathematicians probably do not think that they are working with completed infinite sets of rational numbers. In their hearts, they know that the real numbers are the infinite decimals that are the coordinates of points on a line. And these points are "mathematical" points, points of zero length, width, and height, exact locations in space.

Therefore, the mathematician may routinely work with such

peculiar entities as *half-open intervals*. A half-open interval is a line segment with one (and exactly one) endpoint missing. As with Pegasus, it is instructive to contemplate the existence or nonexistence of such entities, for no one has ever seen one, although they are much discussed.

HEAPS OF SAND

The Greek philosopher Eublides proposed that there is no such thing as a heap of sand. For certainly one grain of sand does not constitute a heap. Adding another grain does not make a heap either, for two grains of sand would not be called a heap. In fact, if you do not have a heap of sand already, adding one grain to what you have will not produce a heap. So you can never get to a heap.

Zeno made essentially the same argument about mathematical points, the kind of points that make up a continuum. He noted that if a point has no size, then adding another point to it would produce a result that was no larger than what you had to start with. Therefore, you could never get to any object of any size at all by building it from mathematical points.

He then turned the argument around to say that if a point does have size, then, since there are an infinite number of them in a line segment, all line segments must be infinitely long. (This argument has been considerably paraphrased for clarity. Zeno's original version is expressed without reference to line segments.)

He also reasoned from the opposite direction. If the line segment can be infinitely divided, then the points that make it up must have no size. However, since the line segment has a size, it cannot be made from points of no size. (Actually, historians are not sure that this version is due to Zeno, but it sounds so much like him that it probably is.)

Related to this general argument is the following question by Zeno:

"A grain of millet falling makes no sound; how can a bushel therefore make a sound?"

This sounds more like a Zen kōan than a paradox. (The archetypical Zen kōan is "What is the sound of one hand clapping?")

All of the above are basically equivalent to the idea that you can never have a heap of sand. Yet, you can go to the beach and pick up a handful of sand and make a heap. You can also take an infinite sequence of rational numbers to get closer and closer to $\sqrt{2}$, which is not a rational number. No number in the sequence is irrational, but $\sqrt{2}$ is irrational. This is a motivation for defining $\sqrt{2}$ as a completed infinite set of rational numbers, as Dedekind did.

Or, for a similar example that appeals to the geometric intuition, if you continue to add sides to a regular polygon, yet get something that comes very close to a circle. Start with an equilateral triangle, proceed to a square, to a regular pentagon, hexagon, and so forth. At what point do you get to the circle? Should a circle be defined as a completed infinity of polygons?

In fact, a regular polygon of n sides, no matter how great n is, is not a circle. However, by using the idea that it gets very close to a circle, you can calculate the circumference and the area of a circle as close as you please.

Although this concept works for circles, a very similar one fails for curved surfaces according to Schwarz's paradox. The curved surface of a cylinder has a finite area, but it can be separated into triangles that, when their number is allowed to increase to infinity, have as the sum of their areas an infinite number.

When you looked at the sum of any specific infinite series in Chapter 2, the series did not have that particular sum for any finite number of terms. However, mathematicians routinely find the sums for an infinite number of terms and use these results in calculation. In Schwarz's paradox, the sum is always finite for a finite number of triangles.

All this suggests that Zeno is talking about the fundamental nature of mathematics. In mathematics, you cannot get a heap by adding one grain at a time, but if you add an infinite number of grains, you have a heap. Think how these arguments apply to the half-open interval—a line segment that has a point missing at one end. Such a figure has a length—say one unit. If you add the missing point to its proper place, the length does not change. Add one grain to a heap, and you still have a heap.

Zeno's most famous paradoxes are those that were discussed by Aristotle. They are so well known to mathematicians and philosophers that each has its own name.

THE DICHOTOMY

The Dichotomy paradox concerns a traveler who is to walk a certain distance. First, he or she must walk half the distance, then half of what remains, then half of what remains, and so forth. It is clear that there will always be half of the remaining distance to go to any point during the walk. Thus, the person can never complete the walk.

If you represent the distance by a line segment one unit long, then The Dichotomy represents the same situation as the following.

Here the labels are not the coordinates of the points, but an indication of the distance remaining. As you can see, the distance remaining is always $1/2^n$ after n parts of the walk. While $1/2$, $1/4$, $1/8$, ..., $1/2^n$, ... has the limit 0, this limit is achieved only after an infinite number of parts of the walk, so the problem reduces to the difficulties Zeno had with the notion of an infinite number of points on a line discussed earlier.

Recall, however, that a version of this same paradox was used to explain the sum of an infinite series (in Chapter 2).

Physical considerations suggest that the sum of an infinite series be defined as the limit of the sequence obtained by summing the terms of the series one more each time. Here is one example of a physical situation that suggests this definition. Say that a tortoise is traveling down the road at a speed of 1 kilometer per hour. At the end of the first hour, he travels 1 kilometer; at the end of the next half hour, he travels 1-1/2 kilometers; at the end of the next quarter, 1-3/4 kilometers, and so forth. The tortoise's progress is represented by the series

$$1 + \frac{1}{2} + \frac{1}{4} + \cdots + \left(\frac{1}{2}\right)^n + \cdots .$$

While this is an infinite series, it is clear to most folks that at the end of 2 hours the tortoise travels 2 kilometers. And 2 is the limit of the sequence of distances as well as the sequence of total times. Therefore, one can write

$$1 + \frac{1}{2} + \frac{1}{4} + \cdots + \left(\frac{1}{2}\right)^n + \cdots = 2.$$

Notice that the "most folks" to whom this argument is clear do not necessarily include Zeno.

Nevertheless, most mathematicians take the statement that the sum of the times is the sum of an infinite sequence and the sum of the distances is the sum of an infinite sequence to imply that the person (or tortoise) can complete the journey. They feel that there is no paradox involved from a mathematical standpoint, since both series have sums. In other words, the mathematics gives the right result, so The Dichotomy is explained.

There is a story about a drill sergeant who was given the difficult task of telling Private Abernathy that his mother had just died. The sergeant was warned to break it to Abernathy gently. So when he had his platoon lined up at attention, the sergeant commanded, "All those men whose mothers are still living take one step forward." Everyone in the platoon stepped forward and the sergeant shouted, "No so goddamn fast, Abernathy!"

Not so goddamn fast, mathematicians.

In commenting on The Dichotomy, Aristotle pointed out that if you count each part of the journey, then you never reach the end of the counting. That is, the parts can be put into a one-to-one correspondence with the natural numbers:

This leads to a stronger form of The Dichotomy (adapted from José Benardete).

Build an electrical machine that moves a pointer along a bar. In the first half unit of time, say 1/2-second, it moves the pointer halfway along the bar. Then it stops to recharge for exactly as long as it has operated—1/2-second. The new charge is enough to cause it to move the pointer 1/4 the length of the bar in 1/4-second. But it must stop again to recharge for 1/4-second before it can continue. Then it moves the pointer 1/8 the length of the bar in 1/8-second. Continuing in this fashion, when does it get the pointer to the other end of the bar?

Here the continuous motion of the traveler is replaced by a discontinuous motion, so Aristotle's comment applies with full force. The machine never stops, so it never reaches the end of its operation.

There is a discussion below of how a physicist might view such a machine, but you could ponder that notion now on your own.

Benardete has also invented a version of The Dichotomy that does not involve motion. Suppose that you have a book containing a mystery story that has been constructed in an unusual way. When you open the book to read the story, you discover that the first page is 1/2-centimeter thick (rather thick for a page, but this is an unusual book). As you read the story, you notice the second page is only 1/4-centimeter thick, the third is 1/8-centimeter thick, and so forth. After you have read part of the story, you lose interest in finishing, but you still want to find out how the mystery is solved. So you turn the book over to look at the last page. When you open the back cover, however, there is nothing to be seen, for the book has no last page.

Consider also the *Thomson lamp*, the thought experiment proposed by James F. Thomson. The Thomson lamp is programmed to go on and off in a pattern similar to the movements of the machine that moves the pointer along the bar. It is on for 1/2-minute, then off for 1/4-minute, on for 1/8-minute, off for 1/16-minute, and so forth. Now suppose that you turn on a Thomson lamp, and at the end of one minute *exactly* you press a switch that keeps the lamp in whatever state it is in at that time. Will it be on or off?

ACHILLES AND THE TORTOISE

Since Achilles is noted for his swiftness and the tortoise for his slowness, the tortoise is given a head start when they race each other. Zeno argues that Achilles first must reach the point where the tortoise was initially. By then, the tortoise will have moved beyond that point. Now the situation is the same as it was at the start of the race. The tortoise has a head start on Achilles. Achilles must again reach the point where the tortoise was (when Achilles reached the point where the tortoise got his first head start). But by the time he arrived, the tortoise has moved on. Now the situation is the same as it was at the start of the race. The tortoise has a head start on Achilles. And so forth. Achilles can never catch, much less pass, the tortoise, so the tortoise wins the race.

This paradox is often formulated with specific distances and speeds given. Say that the original head start for the tortoise is 100 meters and that Achilles can run 10 times as fast the tortoise (to make the numbers come out nice).

In that case, by the time Achilles reaches the tortoise's starting point, the tortoise has moved 10 meters from that point. By the time Achilles reaches the 110-meter point of the race, the tortoise will be at 111 meters. When Achilles is at 111 meters, the tortoise is at 111.1 meters, when Achilles reaches 111.1, the tortoise is at 111.11, and so forth. Clearly, the tortoise can always keep ahead of Achilles.

Not so goddamn fast, Zeno. As the race progresses in this fashion, the tortoise is heading for the point at *exactly* $111.111 \cdots$ meters. When they both get to that point (which they will, since the time intervals are decreasing to 0), Achilles will catch the tortoise at $111.111 \cdots$ meters, pass him in the next instant, and go on to win the race. In fact, you may recognize $111.111 \cdots$ as a rational number—111-1/9. So, the conventional mathematician's explanation of Zeno's paradox is that, since the time intervals decrease to 0 and the sums of both series of distances exist and are each 111-1/9, there is no paradox at all.

Suppose, however, that the time intervals did *not* decrease to 0 as the space intervals went to 0. This could happen in a couple of ways according to relativity theory.

As was noted at the end of Chapter 7, gravitational force causes time to slow down. If gravitational force is much higher for Achilles than it is for the tortoise, then time would go a little slower for Achilles. And, if the gravitational force grew larger as Achilles moved, then the time intervals might not decrease to 0 in step with the space intervals.

There is a concept in physics known as the *black hole*. Most astronomers believe that the black hole is also a concept in reality. One of the chief things you notice about a black hole is that its gravitational force gets much stronger as you approach the black hole. In fact, the gravitational force gets so strong that, within a certain distance of the black hole, nothing can escape from the region. Even light cannot escape. This capture of light by gravity is what makes the hole black.

If the whole earth could be shrunk down to a mathematical point, but still have the same mass, it would form a black hole. As you approached the point that had been the earth, gravity would rise to much more than it had been at the surface of the unshrunk earth (because gravity depends on both mass and distance). Within about 10 centimeters of the point that had been the earth, nothing, not even light, could escape. The sphere around the black hole from which nothing can escape is called the *event horizon*. The event horizon of the one-earth-mass black hole would have a radius of 10 centimeters.

Astronomers suspect that there are black holes in space that have a mass equal to about 10 suns concentrated at a point. If such black holes exist, their event horizons would be about 30 kilometers from the mass point. Now conduct a race between Achilles and the tortoise as a thought experiment, but hold the race in the vicinity of one of those 10-solar-mass black holes. As before, Achilles will give the tortoise a head start. In this version, Achilles will wait in orbit about as far from the black hole as the earth is from the sun. The tortoise is in the same orbit as Achilles, but on the other side of the black hole (which makes the head start about 185 million miles). The following calculations will be mostly borrowed from Harry L. Shipman's book *Black Holes, Quasars, and the Universe*. To use Dr. Shipman's numbers, Achilles' spaceship travels at a speed that will bring the ship to the event horizon in exactly 205 hours. The tortoise, on the other hand, will be assumed to be moving away from Achilles at the same speed a tortoise travels on earth. This speed is so slow that you can treat the tortoise as if it were not moving at all.

(NOTE: In what follows, the fact that both Achilles and the tortoise are moving in orbit can be ignored.)

If the black hole were not in the way, Achilles would pass the tortoise in just about 410 hours: that's 205 hours to the black hole and 205 more hours to the tortoise. But with the black hole be-tween them, the tortoise will win any race in which Achilles comes close enough to the black hole.

Since this is a thought experiment, let us assume that the strong force of gravity near the black hole will not destroy Achilles and his space ship. (In real life, it would.) Let us also assume that Achilles knows better than to go inside the event horizon, from which he could not escape. Instead, he plans to graze the event horizon, which will give him the shortest path (outside of a straight line) to where the tortoise is.

The race starts. It proceeds as might be expected for the first 204 hours, 30 minutes. Achilles is catching up fast; the tortoise is making no headway at all. Now, however, Achilles is getting close to the black hole. According to the tortoise's clock, Achilles reaches a point just 270 kilometers from the event horizon of the black hole in 204 hours, 33 minutes, 50.1129 seconds. But for Achilles, time is beginning to slow down, so his clock on the spaceship says that he has traveled 204 hours, 33 minutes, 49.6681 seconds. (Note that since Achilles' time is becoming slower than the tortoise's time, the tortoise perceives Achilles as having taken longer to reach this point than Achilles himself perceives the time.)

As Achilles gets closer to the event horizon, the time slows down even more. At 150 kilometers from the event horizon, the tortoise's clock reads 204 hours, 33 minutes, 50.1141 seconds, but Achilles' clock reads, 204 hours, 33 minutes, 49.6692 seconds. The difference between the clocks has gone from 0.4448 second to 0.4449 second as Achilles' spaceship traveled just 120 kilometers. An increase of one-ten-thousandth of a second in such a small interval of the race is a portent. At 3 kilometers from the event horizon, the difference is 0.445609 second. At 30 meters from the event horizon, the dif-ference between the tortoise's clock and Achilles' clock is 0.446077 second.

But Achilles' clock is slowing down much faster than these dif-ferences would seem to indicate. At 3 kilometers from the event horizon, each tick of Achilles' clock is matched by more than 3 ticks of the tortoise's clock. By the time Achilles is 30 meters from the event horizon, one second on Achilles' clock matches more than half a minute on the tortoise's clock. If Achilles just grazes the edge of the event horizon, his clock ticks off one second while the tortoise's clock ticks away for thousands of years. When the tortoise's clock shows exactly 205 hours, Achilles would be within 3×10^{-8283} centimeters of the event horizon.   But the tortoise would never see Achilles actually touch the event horizon, for the ratio of the times on their two clocks is going to infinity. Achilles, from the tortoise's point of view, remains near the event horizon forever. From Achilles' point of view, of course, he is near the event horizon for less than a second.

Obviously, the tortoise wins the race.

It is not necessary to imagine such a strange phenomenon as a black hole for the time intervals in the race not to decrease. Suppose that the tortoise's head start is very great and that Achilles' speed can become very great.   Then the time dilation due to special relativity enables the tortoise to win the race.

By special relativity, the formula that relates the tortoise's time, t, to Achilles' time, A_t, and velocity, A_v, is

$$t = \frac{A_t}{\sqrt{1 - (A_v^2/c^2)}},$$

where c is the speed of light in a vacuum.   By measuring velocity in terms of c (i.e., let $c = 1$), the formula simplifies to

$$t = \frac{A_t}{\sqrt{1 - A_v^2}}$$

To make the calculations easier, assume that you can pick up Achilles as his speed reaches fractions of the speed of light that are in the following sequence:

$$A_v = \sqrt{\frac{1}{2}}, \sqrt{\frac{2}{3}}, \sqrt{\frac{3}{4}}, \cdots, \sqrt{\frac{n}{n+1}}, \cdots$$

It is easy to see that this sequence has a limit of 1.

The denominator in the formula also forms a simple sequence when A_v has these values (which is a good reason to choose these values).

$$\sqrt{1-A_v^2} = \sqrt{\frac{1}{2}}, \sqrt{\frac{1}{3}}, \sqrt{\frac{1}{4}}, \cdots, \sqrt{\frac{1}{n}}, \cdots$$

The sequence for $\sqrt{1-A_v^2}$ has a limit of 0. But this means that t goes to infinity as A_v goes to 1. In other words, as Achilles goes faster and faster, the tortoise sees him slow down. In fact, when A_v is $\sqrt{3/4}$, then $\sqrt{1-A_v^2}$ is $\sqrt{1/4} = 1/2$, so

$$t = \frac{A_t}{\frac{1}{2}}$$

at that point. If Achilles runs at that speed for 1 second, the tortoise perceives Achilles approaching for 2 seconds. Time passes twice as fast for the tortoise as it does for Achilles when Achilles' speed is $\sqrt{3/4}$. Using similar reasoning, you can show that, if m is any natural number, when Achilles' speed is $\sqrt{(m^2-1)/m^2}$, then Achilles' time is m times as slow as the tortoise's time.

Will Achilles be able to pass the tortoise?

Actually, you do not have enough information to tell the answer to that question. You need to know how much of a head start the tortoise gets and how fast Achilles is accelerating. Suppose that Achilles can accelerate about 10 meters a second. A year is about 31,560,000 seconds long, and the speed of light is about 300,000,000 meters per second. Therefore, Achilles would reach the speed of light near the end of a year, except that it is impossible for various reasons to reach or pass the actual speed of light. But Achilles would be close enough to the speed of light for time to begin to slow down appreciably. Achilles' time would reach half the tortoise's time at about 22,500,000 seconds. Thus, Achilles' 260th day of the race would appear to be 48 hours long (approximately) to the tortoise. At that time also, it would appear to the tortoise that Achilles' acceleration had slowed to 5 meters per second (although the cause of this is that the tortoise is seeing an acceleration of 10 meters per 2 seconds). Achilles, continuing on, would find that he had reached 97 percent of the speed of light on the 337th day of the race. But by then the tortoise would find Achilles accelerating only one-third as fast, because the tortoise's time is three times as fast as Achilles' time. Before a year was up (on the 347th day, to be more precise), the tortoise would see Achilles grind to a halt.

The head start that the tortoise would need, however, would have to be very large for Achilles not to catch it before the 347th day. However, it should be clear that since Achilles is always traveling less than the speed of light, a head start of one light year should be sufficient.

Of all of Zeno's paradoxes, Achilles and the Tortoise has been the most "popular," probably because it is the only one known to have characters in it. Consequently, others interested in reasoning about the foundations of mathematics have often used Achilles and the Tortoise as characters in dialogs that illustrate their points. This practice originated with Lewis Carroll in his role as logician Charles Lutwidge Dodgson. In an 1895 article in *Mind*, Lewis Carroll's piece "What the Tortoise Said to Achilles" applied a principle similar to one of Zeno's to formal logic. Later logicians picked up the theme, most notably Douglas Hofstadter in his book *Gödel, Escher, Bach: an Eternal Golden Braid*, where Achilles and the Tortoise are featured frequently throughout a 742-page book on the foundations of mathematics and its relationship to artificial intelligence. Hofstadter's version of their discussion of Quine's paradox is given in Chapter 4.

Carroll's paradox in "What the Tortoise Said to Achilles" may be summarized by saying that the Tortoise shows how to stay one thought ahead of Achilles. Suppose that Achilles accepts the following propositions:

A. Things equal to the same are equal to each other.
B. The two sides of this triangle are things that are equal to the same.
Z. The two sides of the triangle are equal to each other.

Furthermore, Achilles asserts that A and B taken together (if true) imply that Z must be true. Carroll's Tortoise is too polite to say, "Not so goddamn fast, Achilles," but he does point out that a rule of inference is needed, which can be proposition C. This rule is "If A and B are true, then Z must be true." But even this does not satisfy the persistent Tortoise. Given A, B and C, you need a rule of inference to derive Z. Call this D. Then given A, B, C, and D, you need a rule. And so forth. Obviously, Z can never be derived, for there must be an infinity of rules needed.

It probably does not matter that rules of inference do not work, since the same line of reasoning can be used to show that there are no axioms or postulates to which they can be applied. For example, suppose that you decide that "Things equal to the same are equal to each other" is an axiom. From the definition of *axiom*, this is an axiom if and only if it is a self-evident truth. In that case, the true axiom should be "It is a self-evident truth that things equal to the same are equal to each other." But this would only be an axiom if it were a self-evident truth, so the really true axiom should be "It is a self-evident truth that it is a self-evident truth that things equal to the same are equal to each other." And so forth.

Since in one case the axioms go away from us (*regress*) in an infinite fashion and in the other the rules of inference also have an infinite regression, logic is impossible. Of course, this result has just been demonstrated by logic.

Here is a paradox of infinite regression that is from Zeno:

Whatever exists is in a place;
Therefore place exists.
So place is in a place, and so forth.

THE ARROW

The Dichotomy and Achilles and the Tortoise assume that the traveler, Achilles, and the Tortoise can at least move. In The Arrow, Zeno argues that motion cannot take place at all.

Consider an arrow in flight. It must occupy a definite place in space that is exactly the same size as the arrow. Therefore, at any instant, it must be stationary. Any object that occupies a place in space equal to itself cannot be in motion. Since this is true of the arrow at each moment in time, it is never in motion.

The Chinese philosopher Huei Shih, who lived in the third century B.C., independently proposed The Arrow along with a number of other almost paradoxical statements. Huei Shih's version, according to the *Chuangtse* (one of the basic sources of Taoism), was "There is a point in time when the head of a flying arrow neither moves nor stops." Another statement of Huei Shih is closer in spirit to Zeno's Arrow paradox, however: "The shadow of a flying bird never moves." Huei Shih also came close to The Dichotomy, but did not quite make it. "Take a stick one foot long and cut it in half every day and you will never come to the end, even after ten thousand years." The *Chuangtse* makes it clear that Huei Shih was trying to "bait" the logicians of his day, which may very well be what Zeno had in mind as well. Many of Huei Shih's pronouncements are merely riddles in statement form, but some are quite provocative: "The eye does not see," for example, is to be understood as true because the brain does the actual seeing. Another riddle and answer is "The finger does not point to a thing, but rather points endlessly beyond it."

The *Chuangtse*'s comments on Huei Shih might well apply to all those who busy themselves with paradoxes: "From the point of view of the universe, Huei Shih's intellectual exercises are but the activities of a humming mosquito or a buzzing gadfly? Of what use are his teachings to the world?"

Zeno's Arrow paradox is not so close to the heart of mathematics as The Dichotomy, so it has received less attention from mathematicians. The mathematician views The Arrow as just another case of Zeno's failing to understand the continuum. For, given the tension in the bow, the angle the bow makes with the ground before the arrow is released, and other information of that sort, the mathematician can write an equation that describes the motion of the arrow. If air resistance and the curvature of the earth are ignored, the equation will be of the form $y = ax^2 + bx + c$, where a, b, and c are numbers determined by the initial conditions. For any number x, representing the distance along the ground from the bowsman, the position of the arrow will be (x, y). If x moves along the continuum of real numbers, then the arrow moves through space.

The mathematician's view depends again on the reality of the continuum. But what physicists have found out about the real world casts great doubt on that point of view.

In 1900, Max Planck discovered—somewhat to his regret—that energy is not continuous. Instead of being available along the continuum (in any amount one could calculate), energy is only available in small packets—quanta. Furthermore, 13 years later Niels Bohr showed that electrons moved from orbit to orbit without apparently passing through the points between the orbits. (Bohr's original explanation has been much modified since then, but the basic point that the subground of matter and space is *discrete*—that is, chunked—rather than continuous has not changed.)

Now at the subatomic level the arrow is a mass of electrons, protons, and neutrons. Physicists are not sure what those entities *really* are, although they know that they sometimes behave like particles and they sometimes behave like waves. But they, and all the energy they contain, are quantized—in discrete, chunked packets of a certain, very small size. It is hard to go from the level of the subatomic particles to the level of the arrow. It seems reasonable, however, to say that if such a thing as an instant of time exists, then at that instant the arrow might well be stationary. At the next instant, it might well be in another place, just as Bohr's electrons go from orbit to orbit. This depends upon time not being quantized also. But suppose that it is?

In any case, relativity theory makes it clear that you cannot tell *in general* what an instant of time is. Just as you saw in looking at Achilles and the Tortoise racing in space, each moving frame of reference carries its own time with it. If the arrow is moving relative to you, its time is different (only slightly, though) from your time.

Now return to the electrical machine whose pointer moves in decreasing amounts for decreasing amounts of time. How does this machine version of The Dichotomy fit in with modern physics? Recall that the machine would never stop if continuum exists. But when you look at quantum considerations, it is apparent that eventually the pointer would move along the bar a distance that is less than a quantum. Since it could not do that, it would stop before an infinity of motions had been completed.

Furthermore, the machine would stop before that because it operates on electricity. The smallest packet of electricity is the *electron*. The amount of power that is used for each move would become smaller with each move. When the power became less than a single electron could supply, the machine would fail to operate as described. Thus, the machine would either stop or begin to recharge with the power (*rest mass*) of a single electron for each move. This is a definite amount, so the series of moves after that would be each the same size. As you know, a series of the form $a + a + a + \cdots + a + \cdots$ does not converge for any value of *a* different from 0, so the pointer would soon reach the end of the bar.

THE STADIUM

Apparently, The Stadium originally concerned three columns of marchers in a stadium—rather like the shows put on at halftime by marching bands. Column A stands still, Column B marches to the right, while Column C marches to the left. When you first observe them, they are positioned so:

$$
\begin{array}{cccc}
A & A & A & A \\
B & B & B & B & \\
 & & C & C & C & C
\end{array}
$$

After an instant of time, however, they have moved so that the columns are all aligned (they must be marching very fast).

$$
\begin{array}{cccc}
A & A & A & A \\
B & B & B & B \\
C & C & C & C
\end{array}
$$

Columns B and C are moving at the same rate. However, while B has passed only one of the A's, C has passed two of the B's. Therefore, the instant of time could not have been an instant at all, for C is moving twice as fast with respect to B as it is to A (or as B is to A).

This paradox simply makes no sense mathematically, so mathematicians have not discussed it extensively. If you assume the continuum, then the A's, say, cannot be four adjacent mathematical points (because four points don't make a heap), but that is what is required for one B to pass an A in an instant of time.

On the other hand, if you look at the real world as made up of tiny quanta, then you have an interesting paradox. If there is such a thing as a quantum of time, then in one quantum of time, the B's move one quantum of distance with respect to the A's, but the C's move 2 quanta of distance with respect to the B's. Since the motions are uniform, how can this happen? Perhaps the quantum is just the right size so that the relativity of time sorts everything out. After all, the B's are moving at a different rate relative to the A's than they are relative to the C's.

CONCLUSION

Those are all the paradoxes of Zeno that have come down through history—and some of them are suspect as to origin. Nevertheless, they constitute a remarkable critique of mathematics that has often been undervalued or dismissed with too little attention paid to the subtlety of Zeno's thought. Now that the quantum theory is known, these criticisms have even more relevance than they did before 1900.

Perceptive and rigorous mathematicians have seen the problem even without recourse to quantum theory. In 1934, David Hilbert and Paul Bernays, in their *Foundations of Mathematics* said in a discussion of The Dichotomy that "we are by no means obliged to believe that the mathematical space-time representation of motion is physically significant for arbitrarily small space and time intervals . . . . The situation is similar in all cases where one believes it possible to exhibit directly an infinity as given through experience or perception . . . . Closer examination then shows that an infinity is actually not given to us at all . . . ."

The amazing thing is that mathematics works so well. In fact, the discoveries of quantum theory or the special theory of relativity were all made through extensive use of mathematics that was built on the concept of the continuum.

On the one hand, as mathematicians have resorted with more and more frequency to the use of infinity, they have found that they must step with increasing care to avoid the slippery patches. On the other, the use of completed infinities, such as Dedekind's definition of the real numbers, has been a major way to put mathematics on a firmer basis. Some have tried to resolve this contradiction by declaring mathematics a game for which mathematicians can make any rules they like (so long as they don't make rules that contradict other rules). Probably this interpretation is acceptable in theory, but in reality mathematicians start with a view of how real things work and make up rules that correspond with that view.

Some philosophers have attempted to show what mathematics would be like if either reality were different or mathematics were built on different aspects of reality. For example, suppose that it always happened that when two objects were placed side by side, a third one would automatically appear. Then the addition table would be different.

Zeno recognized early on that the mathematical way of looking at the world and the scientific way of looking at the world produced contradictory results. As mathematics has grown independently (to some degree) over the centuries, it has been necessary again and again to change the rules slightly so that mathematical paradoxes will become mere fallacies. But from the beginning, from Zeno's time, it was clear that mathematics does not correspond exactly to the real world.

Of course, this does not mean that mathematics is in any way not useful in describing or discovering the real world. It certainly is. But what is discovered, as in the case of quanta, may not fit with mathematics. If you assume that an arrow behaves like a collection of mathematical points, you can use mathematics to describe its motion. If you concentrate on the arrow being a finite collection of small packets of energy, none of which can be located at a particular mathematical point, then you are up the creek.

This situation remains an unresolved paradox.

Selected Further Reading

Bell, E. T. *Men of Mathematics*. New York, Simon and Schuster, 1937. All of the best stories about mathematicians, whether true or not.

Cohen, Paul C. *Set Theory and the Continuum Hypothesis*. Menlo Park, California, W. A. Benjamin, 1966. Fairly tough going, but straight from the horse's mouth.

Davis, Martin. *The Undecidable*. Hewlett, NY, Raven Press, 1975. An interesting anthology, which includes a translation of Gödel's 1931 paper; rather high-level mathematics for the most part.

Eves, Howard and Carroll V. Newsome. *An Introduction to the Foundations and Fundamental Concepts of Mathematics*. New York, Rinehart, 1958. A textbook, but quite readable; most of the common fallacies are included as exercises in the appropriate chapters.

Gardner, Martin. *The Scientific American Book of Mathematical Puzzles and Diversions*. New York, Simon and Schuster, 1959. *The Unexpected Hanging and Other Mathematical Diversions*. New York, Simon and Schuster, 1969. And various other publications. . . These collections of columns from *Scientific American* include lucid discussions of several topics related to paradoxes and generally require limited mathematical knowledge.

Kasner, Edward and James Newman. *Mathematics and the Imagination*. New York, Simon and Schuster, 1940. An enjoyable book that was the original inspiration for my writing about paradoxes.

Kleene, Stephen Cole. *Introduction to Metamathematics*. Princeton, NJ, D. Van Nostrand, 1950. While quite technical, the early chapters are fairly easy reading; I treated Kleene as the final authority for most topics.

Kline, Morris. *Mathematics: The Loss of Certainty*. New York, Oxford University Press, 1980. A lucid, historical account for the nonmathematician of what has happened to the foundations of mathematics over the centuries, with a point of view very similar to mine.

Kneebone, G. J. *Mathematical Logic and the Foundations of Mathematics*. New York, Van Nostrand Reinhold, 1963. Kneebone is much easier reading than most formal books on logic.

Hofstadter, Douglas R. *Gödel, Escher, Bach: an Eternal Golden Braid*. New York, Basic Books, 1979. A delight, even if you don't read all of it.

Hughes, Patrick and George Brecht. *Vicious Circles and Infinity: An Anthology of Paradoxes*. New York, Doubleday, 1975. Most of the common and a few of the uncommon paradoxes are reported with little or no comment on them.

Nagel, Ernest and James R. Newman, *Gödel's Proof*, New York, New York University Press, 1956. This is the least technical explanation of Gödel's incompleteness theorem generally available.

Newman, James R. *The World of Mathematics*. New York, Simon and Schuster, 1956. These four volumes include essays and biographies for the general reader, with quite helpful introductory remarks.

Quine, Willard Van Orman. *The Ways of Paradox and Other Essays. Revised and Enlarged Edition*. Cambridge, Harvard University Press, 1976. The first essay, which grew out of a *Scientific American* article, classifies the types of paradox one might encounter and discusses some of them in detail.

Reid, Constance. *Introduction to Higher Mathematics*. New York, Thomas Y. Crowell, 1959. One of the easiest books for the nonmathematician who would like to understand some mathematics beyond that learned in school.

Tietze, Heinrich. *Famous Problems in Mathematics*. Baltimore, Graylock Press, 1965. While Tietze expects the reader to have a European mathematics education, the slightly different slant he gives to familiar problems (to readers of the mathematics literature) is refreshing.

Index

A CATALOG OF SELECTED
DOVER BOOKS
IN ALL FIELDS OF INTEREST

A CATALOG OF SELECTED DOVER
BOOKS IN ALL FIELDS OF INTEREST

CONCERNING THE SPIRITUAL IN ART, Wassily Kandinsky. Pioneering work by father of abstract art. Thoughts on color theory, nature of art. Analysis of earlier masters. 12 illustrations. 80pp. of text. 5⅜ x 8½. 23411-8 Pa. $4.95

ANIMALS: 1,419 Copyright-Free Illustrations of Mammals, Birds, Fish, Insects, etc., Jim Harter (ed.). Clear wood engravings present, in extremely lifelike poses, over 1,000 species of animals. One of the most extensive pictorial sourcebooks of its kind. Captions. Index. 284pp. 9 x 12. 23766-4 Pa. $14.95

CELTIC ART: The Methods of Construction, George Bain. Simple geometric techniques for making Celtic interlacements, spirals, Kells-type initials, animals, humans, etc. Over 500 illustrations. 160pp. 9 x 12. (USO) 22923-8 Pa. $9.95

AN ATLAS OF ANATOMY FOR ARTISTS, Fritz Schider. Most thorough reference work on art anatomy in the world. Hundreds of illustrations, including selections from works by Vesalius, Leonardo, Goya, Ingres, Michelangelo, others. 593 illustrations. 192pp. 7⅛ x 10¼. 20241-0 Pa. $9.95

CELTIC HAND STROKE-BY-STROKE (Irish Half-Uncial from "The Book of Kells"): An Arthur Baker Calligraphy Manual, Arthur Baker. Complete guide to creating each letter of the alphabet in distinctive Celtic manner. Covers hand position, strokes, pens, inks, paper, more. Illustrated. 48pp. 8¼ x 11. 24336-2 Pa. $3.95

EASY ORIGAMI, John Montroll. Charming collection of 32 projects (hat, cup, pelican, piano, swan, many more) specially designed for the novice origami hobbyist. Clearly illustrated easy-to-follow instructions insure that even beginning papercrafters will achieve successful results. 48pp. 8¼ x 11. 27298-2 Pa. $3.50

THE COMPLETE BOOK OF BIRDHOUSE CONSTRUCTION FOR WOODWORKERS, Scott D. Campbell. Detailed instructions, illustrations, tables. Also data on bird habitat and instinct patterns. Bibliography. 3 tables. 63 illustrations in 15 figures. 48pp. 5¼ x 8½. 24407-5 Pa. $2.50

BLOOMINGDALE'S ILLUSTRATED 1886 CATALOG: Fashions, Dry Goods and Housewares, Bloomingdale Brothers. Famed merchants' extremely rare catalog depicting about 1,700 products: clothing, housewares, firearms, dry goods, jewelry, more. Invaluable for dating, identifying vintage items. Also, copyright-free graphics for artists, designers. Co-published with Henry Ford Museum & Greenfield Village. 160pp. 8¼ x 11. 25780-0 Pa. $10.95

HISTORIC COSTUME IN PICTURES, Braun & Schneider. Over 1,450 costumed figures in clearly detailed engravings–from dawn of civilization to end of 19th century. Captions. Many folk costumes. 256pp. 8⅜ x 11¾. 23150-X Pa. $12.95

STICKLEY CRAFTSMAN FURNITURE CATALOGS, Gustav Stickley and L. & J. G. Stickley. Beautiful, functional furniture in two authentic catalogs from 1910. 594 illustrations, including 277 photos, show settles, rockers, armchairs, reclining chairs, bookcases, desks, tables. 183pp. 6½ x 9¼. 23838-5 Pa. $11.95

AMERICAN LOCOMOTIVES IN HISTORIC PHOTOGRAPHS: 1858 to 1949, Ron Ziel (ed.). A rare collection of 126 meticulously detailed official photographs, called "builder portraits," of American locomotives that majestically chronicle the rise of steam locomotive power in America. Introduction. Detailed captions. xi + 129pp. 9 x 12. 27393-8 Pa. $13.95

AMERICA'S LIGHTHOUSES: An Illustrated History, Francis Ross Holland, Jr. Delightfully written, profusely illustrated fact-filled survey of over 200 American lighthouses since 1716. History, anecdotes, technological advances, more. 240pp. 8 x 10¾. 25576-X Pa. $12.95

TOWARDS A NEW ARCHITECTURE, Le Corbusier. Pioneering manifesto by founder of "International School." Technical and aesthetic theories, views of industry, economics, relation of form to function, "mass-production split" and much more. Profusely illustrated. 320pp. 6⅛ x 9¼. (USO) 25023-7 Pa. $9.95

HOW THE OTHER HALF LIVES, Jacob Riis. Famous journalistic record, exposing poverty and degradation of New York slums around 1900, by major social reformer. 100 striking and influential photographs. 233pp. 10 x 7⅞. 22012-5 Pa. $11.95

FRUIT KEY AND TWIG KEY TO TREES AND SHRUBS, William M. Harlow. One of the handiest and most widely used identification aids. Fruit key covers 120 deciduous and evergreen species; twig key 160 deciduous species. Easily used. Over 300 photographs. 126pp. 5⅜ x 8½. 20511-8 Pa. $3.95

COMMON BIRD SONGS, Dr. Donald J. Borror. Songs of 60 most common U.S. birds: robins, sparrows, cardinals, bluejays, finches, more–arranged in order of increasing complexity. Up to 9 variations of songs of each species. Cassette and manual 99911-4 $8.95

ORCHIDS AS HOUSE PLANTS, Rebecca Tyson Northen. Grow cattleyas and many other kinds of orchids–in a window, in a case, or under artificial light. 63 illustrations. 148pp. 5⅜ x 8½. 23261-1 Pa. $5.95

MONSTER MAZES, Dave Phillips. Masterful mazes at four levels of difficulty. Avoid deadly perils and evil creatures to find magical treasures. Solutions for all 32 exciting illustrated puzzles. 48pp. 8¼ x 11. 26005-4 Pa. $2.95

MOZART'S DON GIOVANNI (DOVER OPERA LIBRETTO SERIES), Wolfgang Amadeus Mozart. Introduced and translated by Ellen H. Bleiler. Standard Italian libretto, with complete English translation. Convenient and thoroughly portable–an ideal companion for reading along with a recording or the performance itself. Introduction. List of characters. Plot summary. 121pp. 5¼ x 8½. 24944-1 Pa. $3.95

TECHNICAL MANUAL AND DICTIONARY OF CLASSICAL BALLET, Gail Grant. Defines, explains, comments on steps, movements, poses and concepts. 15-page pictorial section. Basic book for student, viewer. 127pp. 5⅜ x 8½. 21843-0 Pa. $4.95

BRASS INSTRUMENTS: Their History and Development, Anthony Baines. Authoritative, updated survey of the evolution of trumpets, trombones, bugles, cornets, French horns, tubas and other brass wind instruments. Over 140 illustrations and 48 music examples. Corrected and updated by author. New preface. Bibliography. 320pp. 5⅜ x 8½. 27574-4 Pa. $9.95

HOLLYWOOD GLAMOR PORTRAITS, John Kobal (ed.). 145 photos from 1926-49. Harlow, Gable, Bogart, Bacall; 94 stars in all. Full background on photographers, technical aspects. 160pp. 8⅜ x 11¼. 23352-9 Pa. $12.95

MAX AND MORITZ, Wilhelm Busch. Great humor classic in both German and English. Also 10 other works: "Cat and Mouse," "Plisch and Plumm," etc. 216pp. 5⅜ x 8½. 20181-3 Pa. $6.95

THE RAVEN AND OTHER FAVORITE POEMS, Edgar Allan Poe. Over 40 of the author's most memorable poems: "The Bells," "Ulalume," "Israfel," "To Helen," "The Conqueror Worm," "Eldorado," "Annabel Lee," many more. Alphabetic lists of titles and first lines. 64pp. 5³⁄₁₆ x 8¼. 26685-0 Pa. $1.00

PERSONAL MEMOIRS OF U. S. GRANT, Ulysses Simpson Grant. Intelligent, deeply moving firsthand account of Civil War campaigns, considered by many the finest military memoirs ever written. Includes letters, historic photographs, maps and more. 528pp. 6⅛ x 9¼. 28587-1 Pa. $12.95

AMULETS AND SUPERSTITIONS, E. A. Wallis Budge. Comprehensive discourse on origin, powers of amulets in many ancient cultures: Arab, Persian Babylonian, Assyrian, Egyptian, Gnostic, Hebrew, Phoenician, Syriac, etc. Covers cross, swastika, crucifix, seals, rings, stones, etc. 584pp. 5⅜ x 8½. 23573-4 Pa. $12.95

RUSSIAN STORIES/PYCCKNE PACCKA3bl: A Dual-Language Book, edited by Gleb Struve. Twelve tales by such masters as Chekhov, Tolstoy, Dostoevsky, Pushkin, others. Excellent word-for-word English translations on facing pages, plus teaching and study aids, Russian/English vocabulary, biographical/critical introductions, more. 416pp. 5⅜ x 8½. 26244-8 Pa. $9.95

PHILADELPHIA THEN AND NOW: 60 Sites Photographed in the Past and Present, Kenneth Finkel and Susan Oyama. Rare photographs of City Hall, Logan Square, Independence Hall, Betsy Ross House, other landmarks juxtaposed with contemporary views. Captures changing face of historic city. Introduction. Captions. 128pp. 8¼ x 11. 25790-8 Pa. $9.95

AIA ARCHITECTURAL GUIDE TO NASSAU AND SUFFOLK COUNTIES, LONG ISLAND, The American Institute of Architects, Long Island Chapter, and the Society for the Preservation of Long Island Antiquities. Comprehensive, well-researched and generously illustrated volume brings to life over three centuries of Long Island's great architectural heritage. More than 240 photographs with authoritative, extensively detailed captions. 176pp. 8¼ x 11. 26946-9 Pa. $14.95

NORTH AMERICAN INDIAN LIFE: Customs and Traditions of 23 Tribes, Elsie Clews Parsons (ed.). 27 fictionalized essays by noted anthropologists examine religion, customs, government, additional facets of life among the Winnebago, Crow, Zuni, Eskimo, other tribes. 480pp. 6⅛ x 9¼. 27377-6 Pa. $10.95

FRANK LLOYD WRIGHT'S HOLLYHOCK HOUSE, Donald Hoffmann. Lavishly illustrated, carefully documented study of one of Wright's most controversial residential designs. Over 120 photographs, floor plans, elevations, etc. Detailed perceptive text by noted Wright scholar. Index. 128pp. 9¼ x 10¾. 27133-1 Pa. $11.95

THE MALE AND FEMALE FIGURE IN MOTION: 60 Classic Photographic Sequences, Eadweard Muybridge. 60 true-action photographs of men and women walking, running, climbing, bending, turning, etc., reproduced from rare 19th-century masterpiece. vi + 121pp. 9 x 12. 24745-7 Pa. $10.95

1001 QUESTIONS ANSWERED ABOUT THE SEASHORE, N. J. Berrill and Jacquelyn Berrill. Queries answered about dolphins, sea snails, sponges, starfish, fishes, shore birds, many others. Covers appearance, breeding, growth, feeding, much more. 305pp. 5¼ x 8¼. 23366-9 Pa. $8.95

GUIDE TO OWL WATCHING IN NORTH AMERICA, Donald S. Heintzelman. Superb guide offers complete data and descriptions of 19 species: barn owl, screech owl, snowy owl, many more. Expert coverage of owl-watching equipment, conservation, migrations and invasions, etc. Guide to observing sites. 84 illustrations. xiii + 193pp. 5⅜ x 8½. 27344-X Pa. $8.95

MEDICINAL AND OTHER USES OF NORTH AMERICAN PLANTS: A Historical Survey with Special Reference to the Eastern Indian Tribes, Charlotte Erichsen-Brown. Chronological historical citations document 500 years of usage of plants, trees, shrubs native to eastern Canada, northeastern U.S. Also complete identifying information. 343 illustrations. 544pp. 6½ x 9¼. 25951-X Pa. $12.95

STORYBOOK MAZES, Dave Phillips. 23 stories and mazes on two-page spreads: Wizard of Oz, Treasure Island, Robin Hood, etc. Solutions. 64pp. 8¼ x 11. 23628-5 Pa. $2.95

NEGRO FOLK MUSIC, U.S.A., Harold Courlander. Noted folklorist's scholarly yet readable analysis of rich and varied musical tradition. Includes authentic versions of over 40 folk songs. Valuable bibliography and discography. xi + 324pp. 5⅜ x 8½. 27350-4 Pa. $9.95

MOVIE-STAR PORTRAITS OF THE FORTIES, John Kobal (ed.). 163 glamor, studio photos of 106 stars of the 1940s: Rita Hayworth, Ava Gardner, Marlon Brando, Clark Gable, many more. 176pp. 8⅜ x 11¼. 23546-7 Pa. $12.95

BENCHLEY LOST AND FOUND, Robert Benchley. Finest humor from early 30s, about pet peeves, child psychologists, post office and others. Mostly unavailable elsewhere. 73 illustrations by Peter Arno and others. 183pp. 5⅜ x 8½. 22410-4 Pa. $6.95

YEKL and THE IMPORTED BRIDEGROOM AND OTHER STORIES OF YIDDISH NEW YORK, Abraham Cahan. Film Hester Street based on Yekl (1896). Novel, other stories among first about Jewish immigrants on N.Y.'s East Side. 240pp. 5⅜ x 8½. 22427-9 Pa. $6.95

SELECTED POEMS, Walt Whitman. Generous sampling from *Leaves of Grass*. Twenty-four poems include "I Hear America Singing," "Song of the Open Road," "I Sing the Body Electric," "When Lilacs Last in the Dooryard Bloom'd," "O Captain! My Captain!"–all reprinted from an authoritative edition. Lists of titles and first lines. 128pp. 5$\frac{3}{16}$ x 8¼. 26878-0 Pa. $1.00

THE BEST TALES OF HOFFMANN, E. T. A. Hoffmann. 10 of Hoffmann's most important stories: "Nutcracker and the King of Mice," "The Golden Flowerpot," etc. 458pp. 5⅜ x 8½.   21793-0 Pa. $9.95

FROM FETISH TO GOD IN ANCIENT EGYPT, E. A. Wallis Budge. Rich detailed survey of Egyptian conception of "God" and gods, magic, cult of animals, Osiris, more. Also, superb English translations of hymns and legends. 240 illustrations. 545pp. 5⅜ x 8½.   25803-3 Pa. $13.95

FRENCH STORIES/CONTES FRANÇAIS: A Dual-Language Book, Wallace Fowlie. Ten stories by French masters, Voltaire to Camus: "Micromegas" by Voltaire; "The Atheist's Mass" by Balzac; "Minuet" by de Maupassant; "The Guest" by Camus, six more. Excellent English translations on facing pages. Also French-English vocabulary list, exercises, more. 352pp. 5⅜ x 8½.   26443-2 Pa. $9.95

CHICAGO AT THE TURN OF THE CENTURY IN PHOTOGRAPHS: 122 Historic Views from the Collections of the Chicago Historical Society, Larry A. Viskochil. Rare large-format prints offer detailed views of City Hall, State Street, the Loop, Hull House, Union Station, many other landmarks, circa 1904-1913. Introduction. Captions. Maps. 144pp. 9⅜ x 12¼.   24656-6 Pa. $12.95

OLD BROOKLYN IN EARLY PHOTOGRAPHS, 1865-1929, William Lee Younger. Luna Park, Gravesend race track, construction of Grand Army Plaza, moving of Hotel Brighton, etc. 157 previously unpublished photographs. 165pp. 8⅞ x 11¾.
23587-4 Pa. $13.95

THE MYTHS OF THE NORTH AMERICAN INDIANS, Lewis Spence. Rich anthology of the myths and legends of the Algonquins, Iroquois, Pawnees and Sioux, prefaced by an extensive historical and ethnological commentary. 36 illustrations. 480pp. 5⅜ x 8½.   25967-6 Pa. $10.95

AN ENCYCLOPEDIA OF BATTLES: Accounts of Over 1,560 Battles from 1479 B.C. to the Present, David Eggenberger. Essential details of every major battle in recorded history from the first battle of Megiddo in 1479 B.C. to Grenada in 1984. List of Battle Maps. New Appendix covering the years 1967-1984. Index. 99 illustrations. 544pp. 6½ x 9¼.   24913-1 Pa. $16.95

SAILING ALONE AROUND THE WORLD, Captain Joshua Slocum. First man to sail around the world, alone, in small boat. One of great feats of seamanship told in delightful manner. 67 illustrations. 294pp. 5⅜ x 8½.   20326-3 Pa. $6.95

ANARCHISM AND OTHER ESSAYS, Emma Goldman. Powerful, penetrating, prophetic essays on direct action, role of minorities, prison reform, puritan hypocrisy, violence, etc. 271pp. 5⅜ x 8½.   22484-8 Pa. $7.95

MYTHS OF THE HINDUS AND BUDDHISTS, Ananda K. Coomaraswamy and Sister Nivedita. Great stories of the epics; deeds of Krishna, Shiva, taken from puranas, Vedas, folk tales; etc. 32 illustrations. 400pp. 5⅜ x 8½.   21759-0 Pa. $12.95

BEYOND PSYCHOLOGY, Otto Rank. Fear of death, desire of immortality, nature of sexuality, social organization, creativity, according to Rankian system. 291pp. 5⅜ x 8½.
20485-5 Pa. $8.95

A THEOLOGICO-POLITICAL TREATISE, Benedict Spinoza. Also contains unfinished Political Treatise. Great classic on religious liberty, theory of government on common consent. R. Elwes translation. Total of 421pp. 5⅜ x 8½. 20249-6 Pa. $9.95

MY BONDAGE AND MY FREEDOM, Frederick Douglass. Born a slave, Douglass became outspoken force in antislavery movement. The best of Douglass' autobiographies. Graphic description of slave life. 464pp. 5⅜ x 8½. 22457-0 Pa. $8.95

FOLLOWING THE EQUATOR: A Journey Around the World, Mark Twain. Fascinating humorous account of 1897 voyage to Hawaii, Australia, India, New Zealand, etc. Ironic, bemused reports on peoples, customs, climate, flora and fauna, politics, much more. 197 illustrations. 720pp. 5⅜ x 8½. 26113-1 Pa. $15.95

THE PEOPLE CALLED SHAKERS, Edward D. Andrews. Definitive study of Shakers: origins, beliefs, practices, dances, social organization, furniture and crafts, etc. 33 illustrations. 351pp. 5⅜ x 8½. 21081-2 Pa. $8.95

THE MYTHS OF GREECE AND ROME, H. A. Guerber. A classic of mythology, generously illustrated, long prized for its simple, graphic, accurate retelling of the principal myths of Greece and Rome, and for its commentary on their origins and significance. With 64 illustrations by Michelangelo, Raphael, Titian, Rubens, Canova, Bernini and others. 480pp. 5⅜ x 8½. 27584-1 Pa. $9.95

PSYCHOLOGY OF MUSIC, Carl E. Seashore. Classic work discusses music as a medium from psychological viewpoint. Clear treatment of physical acoustics, auditory apparatus, sound perception, development of musical skills, nature of musical feeling, host of other topics. 88 figures. 408pp. 5⅜ x 8½. 21851-1 Pa. $11.95

THE PHILOSOPHY OF HISTORY, Georg W. Hegel. Great classic of Western thought develops concept that history is not chance but rational process, the evolution of freedom. 457pp. 5⅜ x 8½. 20112-0 Pa. $9.95

THE BOOK OF TEA, Kakuzo Okakura. Minor classic of the Orient: entertaining, charming explanation, interpretation of traditional Japanese culture in terms of tea ceremony. 94pp. 5⅜ x 8½. 20070-1 Pa. $3.95

LIFE IN ANCIENT EGYPT, Adolf Erman. Fullest, most thorough, detailed older account with much not in more recent books, domestic life, religion, magic, medicine, commerce, much more. Many illustrations reproduce tomb paintings, carvings, hieroglyphs, etc. 597pp. 5⅜ x 8½. 22632-8 Pa. $12.95

SUNDIALS, Their Theory and Construction, Albert Waugh. Far and away the best, most thorough coverage of ideas, mathematics concerned, types, construction, adjusting anywhere. Simple, nontechnical treatment allows even children to build several of these dials. Over 100 illustrations. 230pp. 5⅜ x 8½. 22947-5 Pa. $8.95

DYNAMICS OF FLUIDS IN POROUS MEDIA, Jacob Bear. For advanced students of ground water hydrology, soil mechanics and physics, drainage and irrigation engineering, and more. 335 illustrations. Exercises, with answers. 784pp. 6⅛ x 9¼. 65675-6 Pa. $19.95

SONGS OF EXPERIENCE: Facsimile Reproduction with 26 Plates in Full Color, William Blake. 26 full-color plates from a rare 1826 edition. Includes "TheTyger," "London," "Holy Thursday," and other poems. Printed text of poems. 48pp. 5¼ x 7. 24636-1 Pa. $4.95

OLD-TIME VIGNETTES IN FULL COLOR, Carol Belanger Grafton (ed.). Over 390 charming, often sentimental illustrations, selected from archives of Victorian graphics–pretty women posing, children playing, food, flowers, kittens and puppies, smiling cherubs, birds and butterflies, much more. All copyright-free. 48pp. 9¼ x 12¼. 27269-9 Pa. $7.95

PERSPECTIVE FOR ARTISTS, Rex Vicat Cole. Depth, perspective of sky and sea, shadows, much more, not usually covered. 391 diagrams, 81 reproductions of drawings and paintings. 279pp. 5⅜ x 8½. 22487-2 Pa. $7.95

DRAWING THE LIVING FIGURE, Joseph Sheppard. Innovative approach to artistic anatomy focuses on specifics of surface anatomy, rather than muscles and bones. Over 170 drawings of live models in front, back and side views, and in widely varying poses. Accompanying diagrams. 177 illustrations. Introduction. Index. 144pp. 8⅜ x11¼. 26723-7 Pa. $8.95

GOTHIC AND OLD ENGLISH ALPHABETS: 100 Complete Fonts, Dan X. Solo. Add power, elegance to posters, signs, other graphics with 100 stunning copyright-free alphabets: Blackstone, Dolbey, Germania, 97 more–including many lower-case, numerals, punctuation marks. 104pp. 8⅛ x 11. 24695-7 Pa. $8.95

HOW TO DO BEADWORK, Mary White. Fundamental book on craft from simple projects to five-bead chains and woven works. 106 illustrations. 142pp. 5⅜ x 8. 20697-1 Pa. $4.95

THE BOOK OF WOOD CARVING, Charles Marshall Sayers. Finest book for beginners discusses fundamentals and offers 34 designs. "Absolutely first rate . . . well thought out and well executed."–E. J. Tangerman. 118pp. 7¾ x 10⅝. 23654-4 Pa. $6.95

ILLUSTRATED CATALOG OF CIVIL WAR MILITARY GOODS: Union Army Weapons, Insignia, Uniform Accessories, and Other Equipment, Schuyler, Hartley, and Graham. Rare, profusely illustrated 1846 catalog includes Union Army uniform and dress regulations, arms and ammunition, coats, insignia, flags, swords, rifles, etc. 226 illustrations. 160pp. 9 x 12. 24939-5 Pa. $10.95

WOMEN'S FASHIONS OF THE EARLY 1900s: An Unabridged Republication of "New York Fashions, 1909," National Cloak & Suit Co. Rare catalog of mail-order fashions documents women's and children's clothing styles shortly after the turn of the century. Captions offer full descriptions, prices. Invaluable resource for fashion, costume historians. Approximately 725 illustrations. 128pp. 8⅜ x 11¼. 27276-1 Pa. $11.95

THE 1912 AND 1915 GUSTAV STICKLEY FURNITURE CATALOGS, Gustav Stickley. With over 200 detailed illustrations and descriptions, these two catalogs are essential reading and reference materials and identification guides for Stickley furniture. Captions cite materials, dimensions and prices. 112pp. 6½ x 9¼. 26676-1 Pa. $9.95

EARLY AMERICAN LOCOMOTIVES, John H. White, Jr. Finest locomotive engravings from early 19th century: historical (1804–74), main-line (after 1870), special, foreign, etc. 147 plates. 142pp. 11⅜ x 8¼. 22772-3 Pa. $10.95

THE TALL SHIPS OF TODAY IN PHOTOGRAPHS, Frank O. Braynard. Lavishly illustrated tribute to nearly 100 majestic contemporary sailing vessels: Amerigo Vespucci, Clearwater, Constitution, Eagle, Mayflower, Sea Cloud, Victory, many more. Authoritative captions provide statistics, background on each ship. 190 black-and-white photographs and illustrations. Introduction. 128pp. 8⅞ x 11¾. 27163-3 Pa. $14.95

EARLY NINETEENTH-CENTURY CRAFTS AND TRADES, Peter Stockham (ed.). Extremely rare 1807 volume describes to youngsters the crafts and trades of the day: brickmaker, weaver, dressmaker, bookbinder, ropemaker, saddler, many more. Quaint prose, charming illustrations for each craft. 20 black-and-white line illustrations. 192pp. 4⅝ x 6. 27293-1 Pa. $4.95

VICTORIAN FASHIONS AND COSTUMES FROM HARPER'S BAZAR, 1867–1898, Stella Blum (ed.). Day costumes, evening wear, sports clothes, shoes, hats, other accessories in over 1,000 detailed engravings. 320pp. 9⅜ x 12¼. 22990-4 Pa. $15.95

GUSTAV STICKLEY, THE CRAFTSMAN, Mary Ann Smith. Superb study surveys broad scope of Stickley's achievement, especially in architecture. Design philosophy, rise and fall of the Craftsman empire, descriptions and floor plans for many Craftsman houses, more. 86 black-and-white halftones. 31 line illustrations. Introduction 208pp. 6½ x 9¼. 27210-9 Pa. $9.95

THE LONG ISLAND RAIL ROAD IN EARLY PHOTOGRAPHS, Ron Ziel. Over 220 rare photos, informative text document origin (1844) and development of rail service on Long Island. Vintage views of early trains, locomotives, stations, passengers, crews, much more. Captions. 8⅞ x 11¾. 26301-0 Pa. $13.95

THE BOOK OF OLD SHIPS: From Egyptian Galleys to Clipper Ships, Henry B. Culver. Superb, authoritative history of sailing vessels, with 80 magnificent line illustrations. Galley, bark, caravel, longship, whaler, many more. Detailed, informative text on each vessel by noted naval historian. Introduction. 256pp. 5⅜ x 8½. 27332-6 Pa. $7.95

TEN BOOKS ON ARCHITECTURE, Vitruvius. The most important book ever written on architecture. Early Roman aesthetics, technology, classical orders, site selection, all other aspects. Morgan translation. 331pp. 5⅜ x 8½. 20645-9 Pa. $8.95

THE HUMAN FIGURE IN MOTION, Eadweard Muybridge. More than 4,500 stopped-action photos, in action series, showing undraped men, women, children jumping, lying down, throwing, sitting, wrestling, carrying, etc. 390pp. 7⅞ x 10⅝. 20204-6 Clothbd. $27.95

TREES OF THE EASTERN AND CENTRAL UNITED STATES AND CANADA, William M. Harlow. Best one-volume guide to 140 trees. Full descriptions, woodlore, range, etc. Over 600 illustrations. Handy size. 288pp. 4½ x 6⅜. 20395-6 Pa. $6.95

SONGS OF WESTERN BIRDS, Dr. Donald J. Borror. Complete song and call repertoire of 60 western species, including flycatchers, juncoes, cactus wrens, many more–includes fully illustrated booklet. Cassette and manual 99913-0 $8.95

GROWING AND USING HERBS AND SPICES, Milo Miloradovich. Versatile handbook provides all the information needed for cultivation and use of all the herbs and spices available in North America. 4 illustrations. Index. Glossary. 236pp. 5⅜ x 8½. 25058-X Pa. $7.95

BIG BOOK OF MAZES AND LABYRINTHS, Walter Shepherd. 50 mazes and labyrinths in all–classical, solid, ripple, and more–in one great volume. Perfect inexpensive puzzler for clever youngsters. Full solutions. 112pp. 8⅛ x 11. 22951-3 Pa. $4.95

PIANO TUNING, J. Cree Fischer. Clearest, best book for beginner, amateur. Simple repairs, raising dropped notes, tuning by easy method of flattened fifths. No previous skills needed. 4 illustrations. 201pp. 5⅜ x 8½. 23267-0 Pa. $6.95

A SOURCE BOOK IN THEATRICAL HISTORY, A. M. Nagler. Contemporary observers on acting, directing, make-up, costuming, stage props, machinery, scene design, from Ancient Greece to Chekhov. 611pp. 5⅜ x 8½. 20515-0 Pa. $12.95

THE COMPLETE NONSENSE OF EDWARD LEAR, Edward Lear. All nonsense limericks, zany alphabets, Owl and Pussycat, songs, nonsense botany, etc., illustrated by Lear. Total of 320pp. 5⅜ x 8½. (USO) 20167-8 Pa. $7.95

VICTORIAN PARLOUR POETRY: An Annotated Anthology, Michael R. Turner. 117 gems by Longfellow, Tennyson, Browning, many lesser-known poets. "The Village Blacksmith," "Curfew Must Not Ring Tonight," "Only a Baby Small," dozens more, often difficult to find elsewhere. Index of poets, titles, first lines. xxiii + 325pp. 5⅝ x 8¼. 27044-0 Pa. $8.95

DUBLINERS, James Joyce. Fifteen stories offer vivid, tightly focused observations of the lives of Dublin's poorer classes. At least one, "The Dead," is considered a masterpiece. Reprinted complete and unabridged from standard edition. 160pp. 5³⁄₁₆ x 8¼. 26870-5 Pa. $1.00

THE HAUNTED MONASTERY and THE CHINESE MAZE MURDERS, Robert van Gulik. Two full novels by van Gulik, set in 7th-century China, continue adventures of Judge Dee and his companions. An evil Taoist monastery, seemingly supernatural events; overgrown topiary maze hides strange crimes. 27 illustrations. 328pp. 5⅜ x 8½. 23502-5 Pa. $8.95

THE BOOK OF THE SACRED MAGIC OF ABRAMELIN THE MAGE, translated by S. MacGregor Mathers. Medieval manuscript of ceremonial magic. Basic document in Aleister Crowley, Golden Dawn groups. 268pp. 5⅜ x 8½. 23211-5 Pa. $9.95

NEW RUSSIAN-ENGLISH AND ENGLISH-RUSSIAN DICTIONARY, M. A. O'Brien. This is a remarkably handy Russian dictionary, containing a surprising amount of information, including over 70,000 entries. 366pp. 4½ x 6⅛. 20208-9 Pa. $9.95

HISTORIC HOMES OF THE AMERICAN PRESIDENTS, Second, Revised Edition, Irvin Haas. A traveler's guide to American Presidential homes, most open to the public, depicting and describing homes occupied by every American President from George Washington to George Bush. With visiting hours, admission charges, travel routes. 175 photographs. Index. 160pp. 8¼ x 11. 26751-2 Pa. $11.95

NEW YORK IN THE FORTIES, Andreas Feininger. 162 brilliant photographs by the well-known photographer, formerly with *Life* magazine. Commuters, shoppers, Times Square at night, much else from city at its peak. Captions by John von Hartz. 181pp. 9¼ x 10¾. 23585-8 Pa. $12.95

INDIAN SIGN LANGUAGE, William Tomkins. Over 525 signs developed by Sioux and other tribes. Written instructions and diagrams. Also 290 pictographs. 111pp. 6⅛ x 9¼. 22029-X Pa. $3.95

ANATOMY: A Complete Guide for Artists, Joseph Sheppard. A master of figure drawing shows artists how to render human anatomy convincingly. Over 460 illustrations. 224pp. 8⅜ x 11¼. 27279-6 Pa. $11.95

MEDIEVAL CALLIGRAPHY: Its History and Technique, Marc Drogin. Spirited history, comprehensive instruction manual covers 13 styles (ca. 4th century thru 15th). Excellent photographs; directions for duplicating medieval techniques with modern tools. 224pp. 8⅜ x 11¼. 26142-5 Pa. $12.95

DRIED FLOWERS: How to Prepare Them, Sarah Whitlock and Martha Rankin. Complete instructions on how to use silica gel, meal and borax, perlite aggregate, sand and borax, glycerine and water to create attractive permanent flower arrangements. 12 illustrations. 32pp. 5⅜ x 8½. 21802-3 Pa. $1.00

EASY-TO-MAKE BIRD FEEDERS FOR WOODWORKERS, Scott D. Campbell. Detailed, simple-to-use guide for designing, constructing, caring for and using feeders. Text, illustrations for 12 classic and contemporary designs. 96pp. 5⅜ x 8½. 25847-5 Pa. $3.95

SCOTTISH WONDER TALES FROM MYTH AND LEGEND, Donald A. Mackenzie. 16 lively tales tell of giants rumbling down mountainsides, of a magic wand that turns stone pillars into warriors, of gods and goddesses, evil hags, powerful forces and more. 240pp. 5⅜ x 8½. 29677-6 Pa. $6.95

THE HISTORY OF UNDERCLOTHES, C. Willett Cunnington and Phyllis Cunnington. Fascinating, well-documented survey covering six centuries of English undergarments, enhanced with over 100 illustrations: 12th-century laced-up bodice, footed long drawers (1795), 19th-century bustles, 19th-century corsets for men, Victorian "bust improvers," much more. 272pp. 5⅜ x 8¼. 27124-2 Pa. $9.95

ARTS AND CRAFTS FURNITURE: The Complete Brooks Catalog of 1912, Brooks Manufacturing Co. Photos and detailed descriptions of more than 150 now very collectible furniture designs from the Arts and Crafts movement depict davenports, settees, buffets, desks, tables, chairs, bedsteads, dressers and more, all built of solid, quarter-sawed oak. Invaluable for students and enthusiasts of antiques, Americana and the decorative arts. 80pp. 6½ x 9¼. 27471-3 Pa. $8.95

HOW WE INVENTED THE AIRPLANE: An Illustrated History, Orville Wright. Fascinating firsthand account covers early experiments, construction of planes and motors, first flights, much more. Introduction and commentary by Fred C. Kelly. 76 photographs. 96pp. 8¼ x 11. 25662-6 Pa. $8.95

THE ARTS OF THE SAILOR: Knotting, Splicing and Ropework, Hervey Garrett Smith. Indispensable shipboard reference covers tools, basic knots and useful hitches; handsewing and canvas work, more. Over 100 illustrations. Delightful reading for sea lovers. 256pp. 5⅜ x 8½. 26440-8 Pa. $7.95

FRANK LLOYD WRIGHT'S FALLINGWATER: The House and Its History, Second, Revised Edition, Donald Hoffmann. A total revision—both in text and illustrations—of the standard document on Fallingwater, the boldest, most personal architectural statement of Wright's mature years, updated with valuable new material from the recently opened Frank Lloyd Wright Archives. "Fascinating"—*The New York Times*. 116 illustrations. 128pp. 9¼ x 10¾. 27430-6 Pa. $12.95

PHOTOGRAPHIC SKETCHBOOK OF THE CIVIL WAR, Alexander Gardner. 100 photos taken on field during the Civil War. Famous shots of Manassas Harper's Ferry, Lincoln, Richmond, slave pens, etc. 244pp. 10⅝ x 8¼. 22731-6 Pa. $9.95

FIVE ACRES AND INDEPENDENCE, Maurice G. Kains. Great back-to-the-land classic explains basics of self-sufficient farming. The one book to get. 95 illustrations. 397pp. 5⅜ x 8½. 20974-1 Pa. $7.95

SONGS OF EASTERN BIRDS, Dr. Donald J. Borror. Songs and calls of 60 species most common to eastern U.S.: warblers, woodpeckers, flycatchers, thrushes, larks, many more in high-quality recording. Cassette and manual 99912-2 $9.95

A MODERN HERBAL, Margaret Grieve. Much the fullest, most exact, most useful compilation of herbal material. Gigantic alphabetical encyclopedia, from aconite to zedoary, gives botanical information, medical properties, folklore, economic uses, much else. Indispensable to serious reader. 161 illustrations. 888pp. 6½ x 9¼. 2-vol. set. (USO) Vol. I: 22798-7 Pa. $9.95
Vol. II: 22799-5 Pa. $9.95

HIDDEN TREASURE MAZE BOOK, Dave Phillips. Solve 34 challenging mazes accompanied by heroic tales of adventure. Evil dragons, people-eating plants, blood-thirsty giants, many more dangerous adversaries lurk at every twist and turn. 34 mazes, stories, solutions. 48pp. 8¼ x 11. 24566-7 Pa. $2.95

LETTERS OF W. A. MOZART, Wolfgang A. Mozart. Remarkable letters show bawdy wit, humor, imagination, musical insights, contemporary musical world; includes some letters from Leopold Mozart. 276pp. 5⅜ x 8½. 22859-2 Pa. $7.95

BASIC PRINCIPLES OF CLASSICAL BALLET, Agrippina Vaganova. Great Russian theoretician, teacher explains methods for teaching classical ballet. 118 illustrations. 175pp. 5⅜ x 8½. 22036-2 Pa. $5.95

THE JUMPING FROG, Mark Twain. Revenge edition. The original story of The Celebrated Jumping Frog of Calaveras County, a hapless French translation, and Twain's hilarious "retranslation" from the French. 12 illustrations. 66pp. 5⅜ x 8½.
22686-7 Pa. $3.95

BEST REMEMBERED POEMS, Martin Gardner (ed.). The 126 poems in this superb collection of 19th- and 20th-century British and American verse range from Shelley's "To a Skylark" to the impassioned "Renascence" of Edna St. Vincent Millay and to Edward Lear's whimsical "The Owl and the Pussycat." 224pp. 5⅜ x 8½.
27165-X Pa. $5.95

COMPLETE SONNETS, William Shakespeare. Over 150 exquisite poems deal with love, friendship, the tyranny of time, beauty's evanescence, death and other themes in language of remarkable power, precision and beauty. Glossary of archaic terms. 80pp. 5 3/16 x 8¼. 26686-9 Pa. $1.00

BODIES IN A BOOKSHOP, R. T. Campbell. Challenging mystery of blackmail and murder with ingenious plot and superbly drawn characters. In the best tradition of British suspense fiction. 192pp. 5⅜ x 8½. 24720-1 Pa. $6.95

THE WIT AND HUMOR OF OSCAR WILDE, Alvin Redman (ed.). More than 1,000 ripostes, paradoxes, wisecracks: Work is the curse of the drinking classes; I can resist everything except temptation; etc. 258pp. 5⅜ x 8½.   20602-5 Pa. $5.95

SHAKESPEARE LEXICON AND QUOTATION DICTIONARY, Alexander Schmidt. Full definitions, locations, shades of meaning in every word in plays and poems. More than 50,000 exact quotations. 1,485pp. 6½ x 9¼. 2-vol. set.
Vol. 1: 22726-X Pa. $17.95
Vol. 2: 22727-8 Pa. $17.95

SELECTED POEMS, Emily Dickinson. Over 100 best-known, best-loved poems by one of America's foremost poets, reprinted from authoritative early editions. No comparable edition at this price. Index of first lines. 64pp. 5³⁄₁₆ x 8¼.
26466-1 Pa. $1.00

CELEBRATED CASES OF JUDGE DEE (DEE GOONG AN), translated by Robert van Gulik. Authentic 18th-century Chinese detective novel; Dee and associates solve three interlocked cases. Led to van Gulik's own stories with same characters. Extensive introduction. 9 illustrations. 237pp. 5⅜ x 8½.   23337-5 Pa. $7.95

THE MALLEUS MALEFICARUM OF KRAMER AND SPRENGER, translated by Montague Summers. Full text of most important witchhunter's "bible," used by both Catholics and Protestants. 278pp. 6⅝ x 10.   22802-9 Pa. $12.95

SPANISH STORIES/CUENTOS ESPAÑOLES: A Dual-Language Book, Angel Flores (ed.). Unique format offers 13 great stories in Spanish by Cervantes, Borges, others. Faithful English translations on facing pages. 352pp. 5⅜ x 8½.
25399-6 Pa. $8.95

THE CHICAGO WORLD'S FAIR OF 1893: A Photographic Record, Stanley Appelbaum (ed.). 128 rare photos show 200 buildings, Beaux-Arts architecture, Midway, original Ferris Wheel, Edison's kinetoscope, more. Architectural emphasis; full text. 116pp. 8¼ x 11.   23990-X Pa. $9.95

OLD QUEENS, N.Y., IN EARLY PHOTOGRAPHS, Vincent F. Seyfried and William Asadorian. Over 160 rare photographs of Maspeth, Jamaica, Jackson Heights, and other areas. Vintage views of DeWitt Clinton mansion, 1939 World's Fair and more. Captions. 192pp. 8⅞ x 11.   26358-4 Pa. $12.95

CAPTURED BY THE INDIANS: 15 Firsthand Accounts, 1750-1870, Frederick Drimmer. Astounding true historical accounts of grisly torture, bloody conflicts, relentless pursuits, miraculous escapes and more, by people who lived to tell the tale. 384pp. 5⅜ x 8½.   24901-8 Pa. $8.95

THE WORLD'S GREAT SPEECHES, Lewis Copeland and Lawrence W. Lamm (eds.). Vast collection of 278 speeches of Greeks to 1970. Powerful and effective models; unique look at history. 842pp. 5⅜ x 8½.   20468-5 Pa. $14.95

THE BOOK OF THE SWORD, Sir Richard F. Burton. Great Victorian scholar/adventurer's eloquent, erudite history of the "queen of weapons"–from prehistory to early Roman Empire. Evolution and development of early swords, variations (sabre, broadsword, cutlass, scimitar, etc.), much more. 336pp. 6⅛ x 9¼.
25434-8 Pa. $9.95

AUTOBIOGRAPHY: The Story of My Experiments with Truth, Mohandas K. Gandhi. Boyhood, legal studies, purification, the growth of the Satyagraha (nonviolent protest) movement. Critical, inspiring work of the man responsible for the freedom of India. 480pp. 5⅜ x 8½. (USO) 24593-4 Pa. $8.95

CELTIC MYTHS AND LEGENDS, T. W. Rolleston. Masterful retelling of Irish and Welsh stories and tales. Cuchulain, King Arthur, Deirdre, the Grail, many more. First paperback edition. 58 full-page illustrations. 512pp. 5⅜ x 8½. 26507-2 Pa. $9.95

THE PRINCIPLES OF PSYCHOLOGY, William James. Famous long course complete, unabridged. Stream of thought, time perception, memory, experimental methods; great work decades ahead of its time. 94 figures. 1,391pp. 5⅜ x 8½. 2-vol. set.
Vol. I: 20381-6 Pa. $13.95
Vol. II: 20382-4 Pa. $14.95

THE WORLD AS WILL AND REPRESENTATION, Arthur Schopenhauer. Definitive English translation of Schopenhauer's life work, correcting more than 1,000 errors, omissions in earlier translations. Translated by E. F. J. Payne. Total of 1,269pp. 5⅜ x 8½. 2-vol. set.
Vol. 1: 21761-2 Pa. $12.95
Vol. 2: 21762-0 Pa. $12.95

MAGIC AND MYSTERY IN TIBET, Madame Alexandra David-Neel. Experiences among lamas, magicians, sages, sorcerers, Bonpa wizards. A true psychic discovery. 32 illustrations. 321pp. 5⅜ x 8½. (USO) 22682-4 Pa. $9.95

THE EGYPTIAN BOOK OF THE DEAD, E. A. Wallis Budge. Complete reproduction of Ani's papyrus, finest ever found. Full hieroglyphic text, interlinear transliteration, word-for-word translation, smooth translation. 533pp. 6½ x 9¼.
21866-X Pa. $11.95

MATHEMATICS FOR THE NONMATHEMATICIAN, Morris Kline. Detailed, college-level treatment of mathematics in cultural and historical context, with numerous exercises. Recommended Reading Lists. Tables. Numerous figures. 641pp. 5⅜ x 8½.
24823-2 Pa. $11.95

THEORY OF WING SECTIONS: Including a Summary of Airfoil Data, Ira H. Abbott and A. E. von Doenhoff. Concise compilation of subsonic aerodynamic characteristics of NACA wing sections, plus description of theory. 350pp. of tables. 693pp. 5⅜ x 8½. 60586-8 Pa. $14.95

THE RIME OF THE ANCIENT MARINER, Gustave Doré, S. T. Coleridge. Doré's finest work; 34 plates capture moods, subtleties of poem. Flawless full-size reproductions printed on facing pages with authoritative text of poem. "Beautiful. Simply beautiful."–*Publisher's Weekly*. 77pp. 9¼ x 12. 22305-1 Pa. $7.95

NORTH AMERICAN INDIAN DESIGNS FOR ARTISTS AND CRAFTSPEOPLE, Eva Wilson. Over 360 authentic copyright-free designs adapted from Navajo blankets, Hopi pottery, Sioux buffalo hides, more. Geometrics, symbolic figures, plant and animal motifs, etc. 128pp. 8⅜ x 11. (EUK) 25341-4 Pa. $8.95

SCULPTURE: Principles and Practice, Louis Slobodkin. Step-by-step approach to clay, plaster, metals, stone; classical and modern. 253 drawings, photos. 255pp. 8⅛ x 11.
22960-2 Pa. $11.95

CATALOG OF DOVER BOOKS

THE INFLUENCE OF SEA POWER UPON HISTORY, 1660–1783, A. T. Mahan. Influential classic of naval history and tactics still used as text in war colleges. First paperback edition. 4 maps. 24 battle plans. 640pp. 5⅜ x 8½. 25509-3 Pa. $14.95

THE STORY OF THE TITANIC AS TOLD BY ITS SURVIVORS, Jack Winocour (ed.). What it was really like. Panic, despair, shocking inefficiency, and a little heroism. More thrilling than any fictional account. 26 illustrations. 320pp. 5⅜ x 8½. 20610-6 Pa. $8.95

FAIRY AND FOLK TALES OF THE IRISH PEASANTRY, William Butler Yeats (ed.). Treasury of 64 tales from the twilight world of Celtic myth and legend: "The Soul Cages," "The Kildare Pooka," "King O'Toole and his Goose," many more. Introduction and Notes by W. B. Yeats. 352pp. 5⅜ x 8½. 26941-8 Pa. $8.95

BUDDHIST MAHAYANA TEXTS, E. B. Cowell and Others (eds.). Superb, accurate translations of basic documents in Mahayana Buddhism, highly important in history of religions. The Buddha-karita of Asvaghosha, Larger Sukhavativyuha, more. 448pp. 5⅜ x 8½. 25552-2 Pa. $12.95

ONE TWO THREE . . . INFINITY: Facts and Speculations of Science, George Gamow. Great physicist's fascinating, readable overview of contemporary science: number theory, relativity, fourth dimension, entropy, genes, atomic structure, much more. 128 illustrations. Index. 352pp. 5⅜ x 8½. 25664-2 Pa. $8.95

ENGINEERING IN HISTORY, Richard Shelton Kirby, et al. Broad, nontechnical survey of history's major technological advances: birth of Greek science, industrial revolution, electricity and applied science, 20th-century automation, much more. 181 illustrations. ". . . excellent . . ."–*Isis*. Bibliography. vii + 530pp. 5⅝ x 8¼. 26412-2 Pa. $14.95

DALÍ ON MODERN ART: The Cuckolds of Antiquated Modern Art, Salvador Dalí. Influential painter skewers modern art and its practitioners. Outrageous evaluations of Picasso, Cézanne, Turner, more. 15 renderings of paintings discussed. 44 calligraphic decorations by Dalí. 96pp. 5⅜ x 8½. (USO) 29220-7 Pa. $4.95

ANTIQUE PLAYING CARDS: A Pictorial History, Henry René D'Allemagne. Over 900 elaborate, decorative images from rare playing cards (14th–20th centuries): Bacchus, death, dancing dogs, hunting scenes, royal coats of arms, players cheating, much more. 96pp. 9¼ x 12¼. 29265-7 Pa. $12.95

MAKING FURNITURE MASTERPIECES: 30 Projects with Measured Drawings, Franklin H. Gottshall. Step-by-step instructions, illustrations for constructing handsome, useful pieces, among them a Sheraton desk, Chippendale chair, Spanish desk, Queen Anne table and a William and Mary dressing mirror. 224pp. 8⅛ x 11¼. 29338-6 Pa. $13.95

THE FOSSIL BOOK: A Record of Prehistoric Life, Patricia V. Rich et al. Profusely illustrated definitive guide covers everything from single-celled organisms and dinosaurs to birds and mammals and the interplay between climate and man. Over 1,500 illustrations. 760pp. 7½ x 10⅛. 29371-8 Pa. $29.95

Prices subject to change without notice.

Available at your book dealer or write for free catalog to Dept. GI, Dover Publications, Inc., 31 East 2nd St., Mineola, N.Y. 11501. Dover publishes more than 500 books each year on science, elementary and advanced mathematics, biology, music, art, literary history, social sciences and other areas.